"This book is stellar in every sense of the wor
anyone interested in the whys and hows of ex
Its richly textured and imaginative chapters
undoubtedly shape outer space ethnography

– Assistant Professor Will

"This excellent volume is a testament to why and how anthropological engagements with outer space have reached cosmic heights. Each contribution overflows with theoretical and methodological innovations. Anthropologists, particularly those who have imagined the cosmos beyond their purview, will find in this volume inspiration for stretching terrestrial and extraterrestrial inquiries."

– Associate Professor Lisa Messeri, *Yale University, USA*

"Interest in social dimensions of outer space has recently proliferated across the fields of science and technology studies, critical infrastructures research and human geography. However, an essential role in these endeavours is taken by anthropology, which provides both key methodological tools for our enquiries as well as the (self-)critical reflection upon our results.

Having myself developed a 'Peripatetic Approach' to studying socio-technical systems within the Space Sector – spanning a variety of places, communities, temporalities – I welcome this book's comprehensive analysis of ethnographic methods and fieldwork experiences. The authors systematically contextualise both the urgent relevance as well as radical nature of these approaches, noting in particular the various strategies and heuristics to untangle the multi-layered realities and arrange them in accessible interconnected narratives.

This book is essential reading to scholars from social sciences and humanities looking to (further) engage with social studies of outer space, as well as natural scientists and engineers, who are curious about the cultural aspects of their work. Combining reflections from a number of earlier careers scholars with epistemological grounding of their research projects, this work also established a baseline pedagogy concerning ethnographic methods for multi-sited, longitudinal studies, which is accessible to students at all levels."

– Dr Matjaz Vidmar, *Co-founder of Social Studies of Outer Space Network; Author of "Innovation Intermediaries and (Final) Frontiers of High-tech;" Institute for the Study of Science, Technology and Innovation, School of Engineering, The University of Edinburgh, UK*

"*Exploring Ethnography of Outer Space: Methods and Perspectives* is a key text for understanding a burgeoning field: the anthropology of outer space. As well as contributing to the empirical and theoretical delimitation of this field of research, the editors stimulate much-needed methodological reflection on ways of developing ethnographical investigations to study human activities related with outer space."

– Professor Perig Pitrou, *CNRS, Maison Française d'Oxford, Team "Anthropology of Life," Collège de France, PSL University, France*

Exploring Ethnography of Outer Space

This book explores new methods and perspectives in the anthropology of outer space. For the past ten years, scholarship of outer space has grown significantly in the social sciences. Now, an international community of anthropologists is starting to produce significant contributions to this work. This is pushing the conversations around the future of humanity, technology, and outer space beyond the realm of speculative theory into concrete challenges to established norms within anthropology. Each chapter in this volume introduces a unique take on what constitutes an ethnographic field in anthropology. They signal a re-imagination of the central concept for the discipline and offer a timely meditation on the shift in anthropology's understanding of fieldwork from its inception until now. The volume consists of eleven ethnographic chapters, plus an introduction by the editors, and two invited responses. Each of the main body chapters presents a distinct approach to situating outer space empirically on Earth. By bringing together emerging and established scholars, this book ultimately posits that an anthropological approach to outer space requires creative approaches to ethnography that are no longer exclusively premised on a co-presence with the people under study. A primer of innovative ethnographies and an ideal companion to courses on methods, this volume will provide students with a body of accessible, contemporary work on futurisms and outer space. In addition, this book will serve as a snapshot of a moment in ethnographically innovative anthropology that will be relevant to a wider academic audience through its exegesis of new methods for the study of distributed communities.

Jenia Gorbanenko is a PhD candidate in Anthropology at University College London specialising in the anthropology of religion in outer space.

David (Jeeva) Jeevendrampillai is a Lecturer at The University of Manchester. He researches planetary belonging and community building in outer space.

Adryon Kozel is a PhD candidate in Anthropology at University College London. They research how space enthusiasts construct potential futures in space, and narratives of what it means for humans to go to space.

All three editors are members of the ERC Advanced Grant ETHNO-ISS, an anthropological study of the International Space Station based at University College London.

Exploring Ethnography of Outer Space

Methods and Perspectives

Edited by Jenia Gorbanenko,
David (Jeeva) Jeevendrampillai, and
Adryon Kozel

Routledge
Taylor & Francis Group
NEW YORK AND LONDON

Designed cover image: "Hero Shot" by Sarah Fortais and Nikolas Ventourakis, taken at the Grant Museum of Zoology's Micrarium, University College London in 2019.

First published 2025
by Routledge
605 Third Avenue, New York, NY 10158

and by Routledge
4 Park Square, Milton Park, Abingdon, Oxon, OX14 4RN

Routledge is an imprint of the Taylor & Francis Group, an informa business

Funded by University College London.

ISBN: 9781032571294 (hbk)
ISBN: 9781032571287 (pbk)
ISBN: 9781003437956 (ebk)

DOI: 10.4324/9781003437956

Typeset in Sabon
by Newgen Publishing UK

To Koshka

Contents

Contributors

Victor Buchli
University College London, UK
Victor Buchli is professor of material culture in the Anthropology Department at University College London and PI of ETHNO-ISS funded by the European Research Council.

Giles Bunch
University College London, UK
Giles Bunch is a social anthropologist, ETHNO-ISS researcher, and PhD candidate at University College London, researching human spaceflight in the European context.

Davide Chinigò
University of Bologna, Italy
Davide Chinigò is a senior assistant professor in the Department of Political and Social Science of the University of Bologna, and research fellow in the Department of Sociology and Social Anthropology, Stellenbosch University. He studies South Africa's engagement with astronomy and space science.

Ceridwen Dovey
Ceridwen Dovey is an Australian writer and filmmaker.

Sarah Fortais
UCL Centre of Outer Space Studies, UK
Sarah Fortais is a Canadian artist who received a Fine Art PhD from the Slade School of Fine Art (2018), where she built spacesuits for animals, borrowed NASA moon rocks, and created strategies to define what it means to call a person or thing 'cool.'

Jenia Gorbanenko
University College London, UK
Jenia Gorbanenko is an anthropologist of religion and science in outer space at University College London, where she is completing her PhD and conducting research as part of the ETHNO-ISS project.

David (Jeeva) Jeevendrampillai
University of Manchester, UK
Dr David Jeevendrampillai a lecturer at the University of Manchester. During the making of this book, he was a senior research fellow with ETHNO-ISS at UCL. He founded and directed the Centre of Outer Space Studies at UCL. His work concerns anthropology of place, community, the future and modernity; he consults to the space industry and government on issues of space culture.

Anne W. Johnson
Universidad Iberoamerica, Mexico
Anne W. Johnson is a professor of anthropology at the Universidad Iberoamericana in Mexico City and studies Mexican engagements with outer space.

Hae-Seo Kim
University of California, Irvine, US
Hae-Seo Kim is a PhD candidate in the Department of Anthropology at University of California, Irvine, and studies South Korea's engagements with outer space.

Karlijn Korpershoek
Jagiellonian University, Poland
Karlijn Korpershoek is a PhD researcher at the Jagiellonian University, Krakow, and explores life in the space-town of Kourou.

Adryon Kozel
University College London, UK
Adryon Kozel is a PhD candidate in anthropology at University College London. As a member of the ETHNO-ISS project, they study the material culture and social energies of the space community. They are a co-director of UCL's Centre for Outer Space Studies.

Hanna Nieber
Max Planck Institute for Social Anthropology, Germany
Hanna Nieber is a research fellow at the Max Planck Institute for Social Anthropology and investigates astronomy-related activities in Madagascar.

Chakad Ojani
Stockholm University, Sweden
Chakad Ojani holds a PhD in social anthropology from the University of Manchester. He is a Pro Futura Scientia fellow at the Department of Social Anthropology, Stockholm University, and at the Swedish Collegium for Advanced Study. His research primarily focuses on the anthropology of infrastructures, environments, and practices of speculation.

Valerie Olson
University of California, Irvine, US
Valerie Olson is an associate professor in the Department of Anthropology at the University of California, Irvine.

Alana Osbourne
Radboud University, UK
Alana Osbourne is an assistant professor of critical theory at Radboud University.

Rowena Potts
Rowena Potts is a filmmaker and media anthropologist, a co-founder of the Archival Futures Collective, and a lecturer in creative practice at the Australian Film Television and Radio School.

Lauren Reid
Freie Universität, Germany
Lauren Reid is an independent curator and doctoral candidate in social and cultural anthropology at the Freie Universität, Berlin with the research project 'Thinking Beyond the Final Frontier: Cosmic Futures in Thailand.'

Denis Sivkov
Moscow School of Social and Economic Sciences, Russia
Denis Sivkov is a lecturer at Moscow School of Social and Economic Sciences, researching amateur space research and exploration, museums of cosmonautics, and cosmism.

Makar Tereshin
University College London, UK
Makar Tereshin is a social anthropologist and PhD candidate at University College London, where he studies the politics and ecology of the Russian space industry.

Peter Timko
Jagiellonian University, Poland
Peter Timko is a PhD researcher at Jagiellonian University in Krakow, Poland; his work focuses on the contemporary new space economy.

Acknowledgements

The editors would like to acknowledge that compiling this volume was enabled by the European Research Council Advanced Grant ETHNO-ISS (grant no. 833135). We would like to thank the members of the team, specifically Giles Bunch, Aaron Parkhurst, and Victor Buchli, who all acted as a guiding council for the volume as a whole, as well as reviewing our contributions within it. We are grateful for conversations with and the support of David Eller, whose idea to have a book on ethnographies of outer space led to this volume being written. We would like to thank Meagan Simpson at Routledge, whose guidance as editor has been constant, warm, and generous. We would like to thank the wider critical space studies community who have encouraged us to pursue this work. This includes many who may not have contributed to the book via authored work, but who have engaged deeply with its ideas at venues such as the Social Studies of Outer Space annual workshops and other conferences. We would like to thank the contributors to this volume all, of whom have endeavoured to produce exciting and meaningful work that allows this collection to have a vibrancy we could have only hoped for. We would like to acknowledge the generosity of the interlocutors whose engagements with critical social science make our discipline possible.

1 Introduction

Jenia Gorbanenko, David (Jeeva) Jeevendrampillai, and Adryon Kozel

This book explores what happens to forms of ethnography, ideas of the field, and the style of anthropology when outer space figures as a central organising frame for research design, practice, and analysis. It reflects a distinct rise in the critical attention paid by anthropologists to the role of the extraterrestrial in social life. This mirrors the rise in outer space interest in society at large, as the commercialisation of space, expansion of space infrastructures, and rising momentum in human spaceflight mark the so-called 'new space age.' Empirically tracing how outer space figures in social relations on Earth requires that we think creatively and flexibly about notions such as time, scale, and place. This creativity reflects a wider trend in anthropological thinking, whereby the discipline increasingly studies social phenomena beyond a focus on geographically bound communities in fieldwork. Thinking through outer space, we contend, can offer new styles and approaches to anthropological inquiry that widen disciplinary practice in novel and, perhaps, unexpected ways. Each of the eleven essays in this collection demonstrates this productive potential of thinking through outer space, whilst capturing and critically engaging with pivotal geopolitical, social, and economic forces as they unfold.

The growing public interest in outer space is reflected in rising critical inquiry across the social disciplines. Recent contributions to the social research in outer space have come from sociology, history, science and technology studies, archaeology, geography, media and culture studies, and others. Here we highlight some important works that may help the reader to contextualise this volume. *The Palgrave Handbook of Society, Culture and Outer Space* (2016), edited by James S. Ormrod and Peter Dickens, provides an excellent review of the literature on outer space in sociology and geography. Historian Alexander Geppert is one of the central figures spearheading the contemporary historiography of space culture. Geppert and his colleagues' publications in history include the trilogy on European Astroculture: *Imagining Outer Space: European*

DOI: 10.4324/9781003437956-1

Astroculture in the Twentieth Century (Geppert 2018a); *Limiting Outer Space: Astroculture after Apollo* (Geppert 2018b); and *Militarizing Outer Space: Astroculture, Dystopia and the Cold War* (Geppert, Brandau, and Siebeneichner 2021). *The Routledge Handbook of Social Studies of Outer Space* (2023), edited by Juan Francisco Salazar and Alice Gorman, is the most recent comprehensive volume to capture the diversity and dynamism of the current conversation within this community of scholars.

There has been a growing number of academic workshops, conferences, and research networks dedicated to the critical study of outer space. For example, in 2017 David (Jeeva) Jeevendrampillai (one of the editors of this book) organised a workshop titled 'Towards an Anthropology of Outer Space' at University College London (UCL), with Geppert as the invited keynote speaker. This led to the establishment of the Centre for Outer Space Studies which continues to host interdisciplinary talks and research projects. The 2018 European Association for the Study of Science and Technology (EASST) conference had a significant number of space-focused panels and conversations. These led to the establishment of the international Social Studies of Outer Space (SSOS) network which brings together scholars from across the social disciplines and has held an annual symposium since 2022. Their active mailing list regularly advertises conference panels, workshops, and events dedicated to outer space. Our book centres the contributions of a distinct anthropology of outer space which has emerged as part of this wider interdisciplinary lineage.

Whilst space anthropology's history and current shape are further outlined in the following section, we wish to note here that anthropological interest in outer space is not new. Anthropologists have been studying human relationships with outer space and celestial bodies from the discipline's early days. For example, James Frazer's study of magic and religion, *The Golden Bough* (1890), contains an analysis of solar deities. In more recent anthropological works, such as Peter Riviere's (1995) analysis of Ye'Kuana dwellings in Northern Brazil and Venezuela, the roof of the hut is described in terms of its relationship with the night sky. In Suzanne Preston Blier's (1987) analysis of Batammaliba dwellings in Togo, the house mirrors the human body in its symbolic arrangement: the grain store is symbolic of the stomach, the door is symbolic of the mouth, and openings are eyes that track the stars. In this book, we contend that what distinguishes space anthropology from this wider anthropological interest in the celestial is the extent to which outer space is foregrounded in research design, analysis, and discussion. We ask, what does anthropology look like if our research projects, methods, and theory are approached with outer space as the context through which one structures their analytics?

Margaret Mead's work represented one of the first such anthropological engagements that posed questions about outer space explicitly. Three days after the first artificial satellite, Sputnik-1, was launched into orbit by the USSR in 1957, Mead began collecting public perceptions of the event in the US (Mead 1958; Price 2020). At the time of Mead's writing and in the decades that followed, outer space remained a marginal interest until a marked increase in publications in recent years. The special issue of *Anthropological Quarterly*, 'Extreme: Limits and Horizons in the Once and Future Cosmos' (2012), edited by Debbora Battaglia, David Valentine, and Valerie Olson, was a key milestone for the development of the anthropology of outer space. Essays in this collection proposed an anthropology that un-Earths, decentres, and unsettles terrestrial theoretical conventions (Valentine, Olson, and Battaglia 2012; Helmreich 2012; Olson and Messeri 2015; Valentine 2017). They posed fundamental questions about how anthropology, re-designed for a society that is deeply 'more-than-terran' (Olson 2023), advances theoretical conventions developed for a society bound by gravity (Parkhurst and Jeevendrampillai 2020; Valentine 2017). This is a central proposition of space anthropology that this collection builds upon with particular attention to the ethnographic method.

One of the methodological implications that arise out of the foregrounding of outer space is the necessity to conceive of approaches to ethnography that are no longer exclusively premised on a co-presence with the people under study (Buchli 2021). This is increasingly true of all ethnographic fieldwork, especially since the COVID-19 pandemic forced researchers to devise alternative fieldwork arrangements. However, the methodological challenges associated with the study of outer space force anthropologists to do this most explicitly. As noted by Victor Buchli, major changes to the ethnographic method were first precipitated by the increasing prevalence of digital ethnography (Buchli 2021), a methodology that challenges conventional notions of 'ethnographic authority' by further pluralising what the anthropologist's claim that 'I was there' can entail (cf. Clifford 1983; see Boellstroff 2008; Boellstroff et al. 2012; Horst and Miller 2012). Our authors contribute to this re-imagination and expansion of what fieldwork entails and offer a timely meditation on the shift in anthropology's understanding of the fieldsite. As they grapple with multi-sited, multi-temporal, and multi-scalar fieldwork, their work forges new paths for anthropological methods and perspectives. This volume synthesises this contribution to the discipline's development.

In the early stages of this project, we organised the book into three thematic sections. Each section outlined a key conceptual framework, namely space infrastructures, cosmologies of outer space, and practices with people, places, and things. As the chapters took shape, it became clear that these categories were limiting the reading of the works. Using outer space as the primary heuristic in an anthropological work troubles traditional analytical categories that it un-Earths (Messeri 2016; Olson and Messeri 2015), including those we had perceived would order this volume. An infrastructural approach, such as in the work of Chakad Ojani on spaceports in subarctic Sweden (Chapter 3), or in Davide Chinigò and Hanna Nieber's chapter on radio observatories in South Africa and Madagascar (Chapter 4), brings all of these conceptual frameworks – infrastructures, cosmologies, people, places, and things – into relation with outer space. This relationality, premised on the idea of being 'more-than-terran' (Olson 2023), is what binds this collection. In her 2023 piece, Olson develops the term 'more-than-terran,' which – akin to 'more-than-human' (Abram 1996; de la Cadena 2014, 2015) – encapsulates the 'shift into modes of perceiving and working that include but do not centre Earth or dominant conceptualisations of space and boundaries' (Olson 2023, 31). A 'more-than-terran' approach, on the one hand, unbinds the empirical fieldsite from traditional geographies of anthropological enquiry, and, on the other, expands the social scientist's conceptual apparatus for analysing social phenomena.

The chapters in the volume are situated in a narrative order, where each consecutive chapter converses with the previous one, bringing out certain perspectives that we wanted to emphasise. However, the volume can be read in a multitude of other ways. In place of the initial three proposed themes for the organisation of the book's chapters, in the final three sections in this introduction – titled 'Emerging futures,' 'Collapsing scales,' and 'Space heuristic' respectively – we show how the essays are linked by threads. Readers can draw on these threads to make their own connections within and between the essays in this collection.

Space anthropology then and now

As we outlined above, one of the first attempts to launch anthropology into outer space was made at the very outset of the space race between the US and the USSR in the 1950s. Between 1955 and 1964, a group of American social scientists, psychologists, and anthropologists, among them Mead, collaborated on the 'Project Man in Space' (Mead 1958; Mead et al. 1958; Price 2020). The project studied American attitudes towards space exploration with the view of spreading awareness and acceptance of the space programme in the US. One of Mead's main contributions to the project was an analysis of a series of interviews in the US that she and her colleagues conducted following the successful launch of Sputnik-1. Some of this research was then presented at a small conference supported by the American Association for the Advancement of Science (cf. Price 2020). Mead argued that space exploration, and the attendant technological developments, would have different receptions and effects across different cultures, and it is the role of anthropologists to capture and critically engage with this (1953, 1956). She was one of the first to anticipate that being attentive to the extraterrestrial enables the anthropologist to explicate social worlds in novel ways.

A further wave of anthropological interest in outer space came in the late 1960s and 1970s. People travelling further than the Earth's orbit and landing on the Moon prompted some to question whether these reports were true or whether it was beneficial for humanity to venture into outer space at all. During this period, despite not always setting out to research people's attitudes to space exploration, anthropologists still often encountered related issues in their fieldwork, thereby describing tensions around space travel. For instance, during her fieldwork, Irina Pozdeyeva interviewed Russian Old Believers (Eastern Orthodox Christians who split from the Russian Orthodox Church after the seventeenth-century schism), who questioned the very possibility of Americans landing on the Moon. Old Believers understood Heaven to be located somewhere above the skies and doubted that astronauts could go beyond Earth's atmosphere by spacecraft. She observed that they were able to reconcile their beliefs about Heaven and spaceflight by concluding that 'God could not allow man to intrude upon divine celestial space, but he could expand the earthly expanse he had given to man' (1994, 40). Similarly, in her ethnography of a mountain village of northern Evia in Greece, Juliet du Boulay described how the local Greek Orthodox Christians concluded that the story of astronauts landing on the Moon could not be true and 'if it is true one ought not to believe it' (2009, 33). Going up into outer space was a violation of their spiritual geography, where the vertical axis should be preserved for communication with God. They agreed that going against this is not a harmless undertaking, and in the interest of their spiritual health, an Orthodox person should not interrogate this possibility. The presence of humans off Earth called forth a cosmological reckoning across human cultures, which drew the attention of some anthropologists of that era and helped them shed light on cosmological orders from a new angle.

In the US, the Moon landings were widely celebrated, but they were also the subject of cultural critique against a backdrop of popular opposition to the Vietnam War, struggles over gender and reproductive rights, and the Civil Rights Movement (Maher 2018; Scott-Heron 1970; Sun Ra 1974). Around the same time, a theoretical dialogue over whether and how anthropology should deal with the establishment of long-term human settlements beyond Earth emerged. Anthropologist and futurist Magorah Maruyama raised the question of how anthropology, as a discipline, should react and respond to the new realities and possibilities afforded by space travel. Between 1970 and 1974, he ran a series of discussions at the American Anthropological Association (AAA) annual meetings and the International Congress of Anthropological and Ethnological Sciences, which culminated in the publication of an edited collection with the enticing title *Cultures Beyond the Earth: The Role of Anthropology in Outer Space* (Muruyama and Harkins 1975). A speculative proposal, Muruyama and Arthur M. Harkins' collection anticipated challenges for anthropologists conducting research on other planets and with extraterrestrial communities. This book challenged conventional scientific approaches to the production of knowledge by drawing on sociocultural fiction as a mode of engaging with the study of the future. The techno-optimistic contributors also hypothesised about the potential of applied anthropological research for informing the creation of off-Earth human settlements and contact with extraterrestrials. As a work that foregrounds the potential of anthropological research in outer space, the book – perhaps, by virtue of its techno-optimism – has, as of yet, received little critical attention.

In the years following the era of 'firsts' – first satellite in 1957, first crewed spaceflight in 1961, first spacewalk in 1965, first Moon landing in 1969 – the impact of the space industry on Earth has continued to expand. In 2023, the global space economy was valued at $546 billion, having grown by $260 billion (91 per cent) over the past decade (*Space Foundation*, 2023). As such, anthropologists have been turning their attention to outer space activities in their own right. Two landmark ethnographies of the European Space Agency (ESA) examined the implication of its projects for the terrestrial economies and communities they are embedded within: Stacia Zabusky's *Launching Europe: An Ethnography of European Cooperation in Space Science* (1995) and Peter Redfield's *Space in the Tropics: From Convicts to Rockets in French Guiana* (2000). Zabusky's and Redfield's works signalled a crucial turn towards ethnographies of communities that participate in or are affected by the practice of space exploration, thereby developing the discipline beyond its initial interest in the reception of space exploration by society at large. Sean T. Mitchell's research exemplifies this, demonstrating how ethnographies of space launch infrastructure, in his case in Brazil, can illuminate politics of inequality and the attendant issues of technology, race, nation-state, and political consciousness at various scales, from local to international (2017).

The publication of Messeri's *Placing Outer Space* (2016) and Olson's *Into the Extreme* (2018) signalled a further expansion of space anthropology's scope. Working with planetary scientists and space life science labs respectively, these monographs trace the practices of relating to the extraterrestrial, making these relationships the very subject of anthropological research. Messeri focuses on the place-making, sensorial, and emotional practices through which far-away planets are made relatable to humans.

Olson writes about the modes of systems thinking that embed the Earth, celestial bodies, the solar system, and the whole of the extreme environment of outer space within a singular ecological framework of interrelated and nested systems. Valentine has engaged similar questions in his examination of the space settlement advocates' proposals for terraforming Mars (2017; see also Jeevendrampillai and Parkhurst 2021). Sociologist Janet Vertesi's ethnographic research at NASA is another important work that has been formative for the field. Through ethnographically tracing how scientists build, operate, and anthropomorphise Martian rovers, Vertesi demonstrates how they create a sense of familiarity and intimacy with the rover and Mars itself (Vertesi 2015; see also 2020). Their theoretical toolkit has been instrumental for the subsequent generation of space anthropologists.

The emergence of sustained extraterrestrial human communities, as anticipated in Maruyama and Harkins' volume (1975), brought about a new challenge for anthropology – the study of inhabitation of Low Earth Orbit (LEO). Whilst anthropologists have yet to travel to space themselves, two methodological solutions to the problem of ethnographic access to near-space human communities have been put into practice by space researchers. The first involves treating the accounts of astronauts and cosmonauts as sources of auto-ethnographic data about life on space stations. Notable examples of this approach include Jack Stuster's applied research on the behavioural issues associated with long-duration spaceflight for NASA, in which he analysed personal journals maintained by astronauts throughout their missions for the purposes of this study (2010, 2016). Battaglia has similarly engaged with cosmonaut diaries (2012) and 'text-artefacts' that describe aeroponic experiments aboard space stations (2017) in her analyses of inter-human and inter-species extraterrestrial sociality.

The second approach is via the study of the terrestrial communities implicated in the operation of space stations in LEO. In this respect, the launch of the International Space Station (ISS) marked an important development for the social study of near-space. The important methodological work done by archaeologists Alice Gorman and Justin Walsh demonstrated how traditional methods can be brought to the digitally mediated study of remote environments, such as the ISS (Gorman 2023; Walsh and Gorman 2021; see also Castaño 2021, 2023). Although not the first long-term extraterrestrial habitat, it is the longest-running and continuously inhabited space station. As such, Buchli suggests, it is not only a place of scientific discovery – it is a place of dwelling (2021). Crucially, however, the ISS is also the first space station to have amassed such a large and international terrestrial community around it, which has yielded more opportunities for ethnographic access. It has prompted methodological innovation in ethnographic examination, using a re-imagined version of 'armchair anthropology' premised not on physical but on ethnographic co-presence (2021). Terrestrial research with communities participating in the life of the ISS treats the ISS as a combined terrestrial and extraterrestrial nexus of social and material relations, and considers the social arrangements that make the ISS possible (Buchli 2021; Jeevendrampillai et al. 2023; Patarin-Jossec 2021). With the anticipated proliferation of new space stations and the establishment of Moon settlements in the coming years, the salience of ethnographic research of extraterrestrial habitats is likely to grow further.

Following the release of seminal works throughout the 2010s, there has been a marked increase in interest in the anthropology of outer space in the 2020s. The research networks and teams, workshops, conferences, and key publications of the last few years demonstrate the scope and form of this field (see also recent reviews from Eller 2022; Salazar and Gorman 2023; Szolucha et al. 2023; Timko et al. 2022). Whilst anthropological research on outer space has been predominantly US-focused at the outset, it is expanding internationally, in particular in Europe. Two European research grants have solidified the European field with a new generation of post-docs and PhDs: the European Research Council-funded ETHNO-ISS group at UCL in the UK (of which the editors of this book are members) and the National Science Centre Poland-funded Anthropological Research into the Imaginaries and Exploration of Space (ARIES) project at the Jagiellonian University in Poland (of which members Karlijn Korpershoek, Chakad Ojani, and Peter Timko have contributed to this volume). The field's vitality and growing potential are further evidenced by the relatively new (at the time of writing) ethnographic project led by Nina Klimburg-Witjes that received funding from the ERC to explore the European Ariane Rocket programme. Members of these projects are also active participants in the wider international community of social scientists of outer space – the SSOS interdisciplinary network, whose inaugural annual conference 'Ethnographies of Outer Space: Methodological Opportunities and Experiments' was hosted at the University of Trento in 2022 and organised by Valentina Marcheselli and Istvan Praet. Many of the contributions to this volume have grown out of the papers presented at this conference in 2022 and the SSOS conference organised by the ETHNO-ISS project at UCL in London in 2023. In archaeology, Gorman and Walsh have an ongoing project dedicated to the archaeological study of the ISS – the International Space Station Archaeological Project (ISSAP). Another key centre for research in space anthropology is the *Anthropology of Life* group led by Perig Petrou at the Laboratoire d'Anthropologie Sociale at the Collège de France, Université Paris Sciences et Lettres (PSL). The *Off-Earth Atlas*, a joint project between PSL and UCL, is a forthcoming volume that will encapsulate this field for a publication aimed at a general reader (see also Petrou and Praet forthcoming). The PSL *Anthropology of Life* team is part of a wider research direction that considers the study of the possibility of life on other planets by astrobiologists, scientists engaged in the search for extraterrestrial intelligence (SETI), and those interested in the phenomenon of UFOs. Here it is also worth highlighting Debbora Battaglia's edited volume *E.T. Culture: Anthropology in Outerspaces* (2006), Katherine Denning's work at the Society for Social and Conceptual Issues in Astrobiology (SSoCIA), William Lempert and the Indigenous Studies Working Group's critical approach addressing the colonial dynamics in the ethics of contact (Atalay et al. 2021), Michael P. Oman-Reagan's work at the SETI Post-Detection Hub, and the panel on anthropological approaches to UFOs convened by Hussein Agrama at the 2023 AAA/CASCA annual meeting. In addition to his important work on Indigenous futurisms and the ethics of space exploration, Lempert has been at the heart of fostering this warm and supportive research community. It is hard to underestimate the importance of the annual (and informal) *Space Hour* drinks that he organises at the AAA conferences. To all these wonderful people, we owe the possibility and also the pleasure of putting this collection together.

We note that there are areas of research in the critical study of space exploration that anthropologists have had relatively little engagement with to date. Although research into the history of the Chinese and Indian national space programmes is expanding (Harvey 2019; Siddiqi forthcoming; Singh 2017), anthropological research on outer space in these countries is still nascent. Where scholars are able to secure access to national space agencies (Aiken 2015; Messeri 2016; Mirmalek 2020; Olson 2018; Reid 2023; Vertesi 2015, 2020), their access is almost exclusively restricted to civilian initiatives. Militarisation of outer space is a known unknown within anthropology of outer space, and here historians (Geppert, Brandau, and Siebeneichner 2021) but also political scientists (Adamsky 2019; Bowen 2022; Deudney 2020; Privalov 2023) have a much greater reach.

Emergent futures

A lot has changed in anthropology and the world that it researches since Sputnik-1 launched in 1957 and Mead laid the foundations for a new field of anthropological inquiry. The first national space agencies and the attendant goals of establishing human colonies on other celestial bodies took shape at around the same time as the global geopolitical models were shifting post-WWII. As Geppert states:

> An entire geography of outer space [...] presented itself as a continuation, if not a logical extension of earlier geographies of imperial expansion

> and colonial domination [...]. At the same time, outer space developed into one of the major sites of 20th-century utopian thinking, where relations vis-à-vis science, technology and the future were positioned, played out and negotiated.
>
> (2018a, 4)

To paraphrase, the space project emerged within a tension between space exploration as a manifestation of past colonial visions of modernity redirected off Earth, and space exploration as a generative field for new more equitable utopian visions of the future. This tension continues to play out across the world to this day as more and more states invest in the development of national space agencies and infrastructure.

Meanwhile, anthropology has undergone its own postcolonial transformations. In particular, with the self-reflexive turn of the 1980s came a substantial challenge to how time was approached in anthropological analysis. This allowed a wider engagement with the future in anthropology. The work of scholars, such as Johannes Fabian (1983) and Nancy Munn (1986; 1992), drew attention to the enduring colonial logics in anthropology's temporal distancing 'between the West and the Rest' (Fabian 1983, 28). This distancing, described by Fabian as a 'denial of coevalness' (1983), meant that anthropology's 'others' tended to be portrayed as frozen in time, relegated to the past, and without much engagement with the future. It limited the study of the future in non-Western contexts (Munn 1992, 115) and also manifested in a lack of critical engagement with the construction of anthropologist's own time as normative (Munn 1992). In critically isolating this phenomenon, the reflexive turn opened the discussion to the widespread anthropological engagement with the future (Valentine and Hassoun 2019) and analysis of anthropology's methodology and conceptual apparatus for the study of time. Zara Mirmalek, in her research with scientists working on Martian exploration, draws attention to how time is also terrestrially contingent. She notes how, when 'working on Mars time brought to the surface unacknowledged assumptions about the relationship of time and work in scientific knowledge making, about creating standards and leaving them unexamined, and about the bodily relationship between humans and sunlight' (2020, 5; see also Bass, Wales, and Shalin 2005). The works in this collection recognise terrestrial contingencies, such as notions of time, and that being attuned to how such notions are constructed requires that ethnography be written in reflexive, and often creative ways.

The reflexive turn, in particular the *Writing Culture* debate (Clifford and Marcus 1986), challenged the methods of ethnographic representation and anthropology's claim to objectivity. It embraced explicit analysis of the ethnographer's positionality and how it impacts their subjectivity in fieldwork accounts, and welcomed experimentation with collaborative modes of ethnography that challenged hierarchies between the researcher and the researched. This constructive attention to ethnographer's subjectivity led to a re-thinking of the ethnographic method, including the explicit incorporation of artistic practice. From Clifford Geertz's literary experimentations with representation of ethnographic knowledge (1988) to more recent use of drawing, weaving, and sculpting as methods of ethnography (Ingold 2011), anthropologists within this lineage draw attention to the elusiveness of the ethnographic present. They argue against the representation of things as overly fixed in time, and propose an anthropology more attentive to 'attunement' (Stewart 2011) and 'emergence' (Maurer 2005) – the study of things in flux and movement (Ingold 2022).

Some of the more experimental essays in this book continue in the spirit of recent works, such as the volume *Anthropologies and Futures: Researching Emerging and Uncertain Worlds* (Salazar et al. 2017), that highlight the potential of artistic modes of ethnography for the study of emerging futures. This approach proves particularly useful for engaging with some of the more nascent space initiatives. Chakad Ojani (Chapter 3) draws attention to the science fictional qualities of anthropology to unsettle the conventionally Earth-bound anthropology of infrastructure in his research in subarctic Sweden, where residents of a mining city are making sense of the increasing role of outer space in their region's economy. Anne W. Johnson (Chapter 7) reflects on how, in becoming part of the 'space milieu' that she researches in Mexico, she actively participated in shaping how the emergent Latin American 'outer space culture' is conceptualised by her interlocutors, who are increasingly participating in the New Space economy. Ethnographic filmmakers Ceridwen Dovey and Rowena Potts (Chapter 8) describe their collaborative project – *Archival Futures of Outer Space Film Quartet* – that re-animates and re-purposes archival footage for ethnographic art-making as a way of illuminating the speculative fabric of human futures in outer space. Anthropologist David (Jeeva) Jeevendrampillai and artist Sarah Fortais (Chapter 9) vacillate between *Mars* and Earth in the performative act of simulating a future Mars mission to develop healthcare protocols in remote and extreme environments. Lauren Reid's essay (Chapter 10) reflects on her use of a co-creative filmmaking workshop as a way of ethnographically situating how different cosmologies might intersect in future extraterrestrial encounters. These creative practices enable our contributors to empirically access future orientations in the process of their emergence.

By attending to futures as they emerge, contributors to our volume foreground the tensions between colonial pasts and space exploration futures; between the empowerment that investment into outer space research affords to national economies and the people that are impacted on a more local scale. In her chapter about the curatorial display at the Royal Observatory in London, Alana Osbourne (Chapter 5) demonstrates how, despite the critical engagement with the British Empire's colonial past, the museum complex's presentation of the interplanetary exploration and mining projects are steeped in colonial rationales. Davide Chinigò and Hanna Nieber (Chapter 4) highlight how visions of space futures can be simultaneously empowering and marginalising. Their research on the radio observatory infrastructure project, demonstrates how it is both a conduit for the dreams of 'African Renaissance' and a physical obstacle for the local economies it displaces. Moreover, in Madagascar – one of their research sites – the mere anticipation of this infrastructural project is already having a tangible social effect, promoting investment in astrophysics education. Chapters within this volume attend critically both to the colonial rationalities that live on in the projects of off-Earth economies, and to the dreams of a 'good life' (Appadurai 2013; Robbins 2013) that fuel them. They advance an anthropology that is both about the contingencies and impacts of past and present imaginations of the future (Abram and Weszklanys 2013; Bryant and Knight 2019; Hastrup and Skrydstrup 2013; Rosenberg and Harding 2005), and the future as it unfolds in the present (Pink and Salazar 2017), often using artistic practice to attune to these processes.

Collapsing scales

Research about outer space is multi-scalar by design (see also Szolucha et al. 2023), as it concurrently deals with the cosmic, planetary, national, and local scales. For example, when writing about the European Space Agency's (ESA) Kourou spaceport in French Guiana in her chapter (Chapter 12), Karlijn Korpershoek brings European activities off Earth into relation with inter-continental politics, national French politics, and local communities around the spaceport. Hers and other ethnographic projects within this collection each present a unique approach to researching relationality across multiple scales, as they emerge from the fieldwork. There are two important methodological challenges to this type of ethnographic production: how to empirically capture relationships constituted by people and things separated by physical distance and operating on different scales, and how to access moments, in which our interlocutors collapse distances and scales in the fabric of everyday life.

In the aforementioned chapter by Chinigò and Nieber (Chapter 4), the authors draw upon the affordances of their respective disciplines – history and anthropology – in order to trace how a radio astronomy infrastructural project is realised through the bringing together of global financial investments, international aspirations for the 'African Renaissance' and continental unity, individual nations' interests, and local communities' support or resistance. Similarly, Hae-Seo Kim (Chapter 11) demonstrates how relationships with celestial bodies via astrology appear to inform national policy decisions by government officials with regard to the construction of South Korean space agency's centres. Astrology, Kim demonstrates, also informs how these policy decisions are interpreted and received by the people living in the cities, where such space centres are located or planned. We suggest that you might also choose to read across chapters to compare how space exploration projects differently reflect the dynamics of colonial histories and existing political power structures in the making of the future. Osbourne, through the prism of the Royal Observatory in London, (Chapter 5), writes about the concealed vestiges of British colonial logics in UK's current aspirations for Moon exploration. This can be fruitfully read alongside Chinigó and Nieber's description of how South Africa, formerly colonised by the British, is using observatory infrastructure as a conduit for its global aspirations and, concurrently, continental dominance (Chapter 4). Similarly, Giles Bunch's chapter (Chapter 2) on his fieldwork at the ESA's European Astronaut Centre in Cologne, Germany, contrasts nicely with Korpershoek's material on the ESA launch site in Kourou, French Guiana (Chapter 12) with regards to people's sense of proximity to space science activities. Peter Timko (also in Chapter 12) describes the experiences of empowered and ambitious students prefiguring the future of the space sector with their university rocket clubs in California. This can be juxtaposed with Johnson's account (Chapter 7) of the challenges experienced by her Mexican interlocutors in their attempts to realise their aspirations for the Mexican Space Agency in spite of the limited funding and access to the space industry. These chapters, when read alongside each other, highlight the importance of perspective and positionality, from which different scales or vast distances are apprehended.

The relationship between the local and the extraterrestrial (Szolucha et al. 2023) is a central concern within this book. As the researcher may turn to ethnographic innovations to capture the relationality between people and things across scales, our ethnographic interlocutors are also drawing relations between scales in novel ways. These ethnographies bring to attention the practices of collapsing scales and the folding of outer space into the present moment. As we are taken on a tour of a small space museum on the margins of the Russian space exploration community, Makar Tereshin and Denis Sivkov (Chapter 6) describe how the universe is brought closer and made sensible to museum visitors. The intimate interaction with the meteorite in the Royal Observatory described by Osbourne (Chapter 5) similarly connects the museum-goers to future imperial projects of extractivism off Earth. At other times, the scales that our contributors' interlocutors are dealing with are cosmological, and, rather than being collapsed, the authors observe how they are woven together by the people inhabiting a pluriverse of multiple cosmologies. Both Reid (Chapter 10) and Kim (Chapter 11) attend to the cosmopolitics of the 'working together of divergent cosmologies' (de la Cadena and Blaser 2018, 12), looking at how Buddhism and Korean astrology intersect with the scientific cosmology of the universe in their respective field sites.

Space heuristic

Outer space offers a unique heuristic in terms of field design, ethnographic style, and analysis to the wider field of anthropology. In their introduction to *The Routledge Handbook of Social Studies of Outer Space,* Salazar and Gorman assert that outer space is not only a site and a research object – it is also an epistemic framework (2023, 6). Space anthropology is both an area studies in a traditional sense: one that researches outer space as a place. It is also a thematic field that approaches outer space as the object of inquiry. But treating outer space as an epistemic framework also brings phenomena into relation with each other in novel ways, by centring outer space as the driving force behind the formation of these social relations. For example, in Osbourne's chapter (Chapter 5), a rock from outer space becomes a portal that shows forth the material link between the British history of extractive colonialism on Earth and present-day aspirations for a space-faring future. Bunch's chapter (Chapter 2), regarding the perceived hierarchies of labour within the ESA, demonstrates how using outer space as a heuristic ties together traditional anthropological work on hierarchy with the anthropology of biomedicine.

Whilst most of the chapters demonstrate how one might foreground outer space in their research, Timko and Korpershoek's chapter, near the end of the volume (Chapter 12), suggests how shifting the focus away from outer space can be a productive exercise in its own right. They show the possibilities opened up by resisting the temptation to overly contain and define outer space, or, as they say, by 'leaving space blank.' We agree that it is precisely its openness, its double meanings, and the entanglement between the universe as a place and as a human cosmology (Abramson and Holbraad 2016) nested within the idea of outer space, that makes outer space such a productive research framework for cohering things into relation (Salazar and Gorman 2023). Even when it is left undefined.

Researching how communities of people relate with outer space creates an empirical entry point into the discussion of anthropological concepts in a post-planetary age (Tabas 2022; Battaglia, Valentine, and Olson 2012; Olson and Messeri 2015; Szolucha et al. 2023). The purpose of this collection is to demonstrate the plurality of how this can be done, and the diversity of research themes and ethnographic approaches within space anthropology. As a primer of recent ethnographies, this volume is an ideal companion to those wishing to research through the heuristic of outer space and those in university courses in anthropology. This is thanks to the conceptual breadth it covers and the practical methodological analysis that it provides. As a snapshot of a moment in ethnographically innovative anthropology, it offers an exegesis of new methods for the study of distributed communities and practices on multiple scales. The methodological challenges posed by the research of outer space and tackled by the authors in this collection have wide-ranging applications in studies of the future, art, and infrastructure, making it relevant for a wide academic audience.

In the early preparatory stage of this volume, we had a conversation with Valerie Olson, who has also kindly agreed to join the project as a respondent to reflect on the collection's methodological contributions. She suggested that, as someone who teaches space anthropology, she needs a book that she can give to burgeoning researchers to show why one should, and how one might, design a project about outer space. We envision that this book will do precisely that: it is the why and how for outer space ethnography. We also invited our second respondent, Victor Buchli, to speak of the extent to which these essays deliver on the theoretical promise of un-Earthing anthropological concepts. We hope it brings you as much joy to read as it did to us to compile. We believe it offers hints and insights into thinking about anthropology in, as Olson (2023) would say, a 'more-than-terran' way.

References

Abram, David. 1996. *The Spell of the Sensuous: Perception and Language in a More-than-Human World*. New York: Vintage Books/Random House.

Abram, Simone and Gina Weszkalnys. 2013. *Elusive Promises: Planning in the Contemporary World*. Oxford and New York: Berghahn.

Abramson, Allen, and Martin Holbraad (eds). 2016. *Framing Cosmologies: The Anthropology of Worlds*. Manchester: Manchester University Press.

Adamsky, Dmitry. 2019. *Russian Nuclear Orthodoxy: Religion, Politics, and Strategy*. Stanford, CA: Stanford University Press.

Aiken, Jo. 2015. "Otherworldly Anthropology: Past, Present, and Future Contributions of Ethnographers to Space Exploration." In *Applied Anthropology: Unexpected Spaces, Topics, and Methods*, edited by Sheena Nahm and Cortney Hughes Rinker. London: Routledge.

Appadurai, Arjun. 2013. *The Future as Cultural Fact: Essays on the Global Condition*. London: Verso.

Atalay, S., W. Lempert, S. D. Shorter, and K. Tallbear. 2021. "Indigenous Studies Working Group Statement." *American Indian Culture and Research Journal* , 45, no. 1: 9–18.

Bass, Deborah S., Roxana C. Wales, and Valerie L. Shalin. 2005. "Choosing Mars-Time: Analysis of the Mars Exploration Rover Experience." Conference Paper, IEEE Aerospace Conference.

Battaglia, Debbora (ed.). 2006. *E.T. Culture: Anthropology in Outerspaces.* Durham, NC: Duke University Press.

Battaglia, Debbora. 2012. "Coming in at an Unusual Angle: Exo-Surprise and the Fieldworking Cosmonaut." *Anthropological Quarterly* 85, no. 4: 1089–1106.

Battaglia, Debbora. 2017. "Aeroponic Gardens and Their Magic: Plants/Persons/ Ethics in Suspension." *History and Anthropology* 28, no. 3: 263–292.

Battaglia, Debbora, David Valentine, and Valerie A. Olson (eds). 2012. "Special Collection. Extreme: Humans at Home in the Cosmos." *Anthropological Quarterly* 85, no. 4.

Blier, Suzanne Preston. 1987. *The Anatomy of Architecture: Ontology and Metaphor in Batammaliba Architectural Expression.* Cambridge: Cambridge University Press.

Boellstorff, Tom. 2008. *Coming of Age in Second Life: An Anthropologist Explores the Virtually Human.* Oxford and Princeton, NJ: Princeton University Press.

Boellstorff, Tom, Bonnie Nardi, Celia Pearce, and T.L. Taylor. 2012. *Ethnography and Virtual Worlds: A Handbook of Method.* Princeton, NJ: Princeton University Press.

Bowen, Bleddyn E. 2020. *War in Space: Strategy, Spacepower, Geopolitics.* Edinburgh: Edinburgh University Press.

Bowen, Bleddyn E. 2022. *Original Sin: Power, Technology and War in Outer Space.* London: Hurst Publishers.

Bryant, Rebecca and Daniel M. Knight. 2019. *The Anthropology of the Future.* Cambridge: Cambridge University Press.

Buchli, Victor. 2021. "Extraterrestrial Methods: Towards an Ethnography of the ISS." In *Lineages and Advancements in Material Culture Studies: Perspectives from UCL Anthropology*, edited by Timothy Carroll, Tone Walford, and Shireen Walton, 17–32. London: Routledge.

Castaño, Paola. 2021. "From Value to Valuation: Pragmatist and Hermeneutic Orientations for Assessing Science on the International Space Station." *The American Sociologist*, 52, no. 4: 671–701. Abstract.

Castaño, Paola. 2023. "Plant Biologists and the International Space Station: Institutionalising a Scientific Community". In *The Routledge Handbook of Social Studies of Outer Space*, edited by Juan Francisco Salazar and Alice Gorman, 423–436. New York: Routledge.

Clifford, James. 1983. "On Ethnographic Authority." *Representations* 1, no. 2 (Spring): 118–146.

Clifford, James and George E. Marcus (eds). 1986. *Writing Culture: the Poetics and Politics of Ethnography.* Berkeley and London: University of California Press.

De la Cadena, Marisol. 2014. "Runa: Human But Not Only." *HAU: Journal of Ethnographic Theory* 4, no. 2: 253–59.https://doi.org/10.14318/hau4.2.013.

De la Cadena, Marisol. 2015. *Earth Beings: Ecologies of Practice Across Andean Worlds*. Durham, NC and London: Duke University Press.

De la Cadena, Marisol and Mario Blaser (eds). 2018. *A World of Many Worlds*. Durham, NC and London: Duke University Press.

Deudney, Daniel. 2020. *Dark Skies: Space Expansionism, Planetary Geopolitics, and the Ends of Humanity*. New York: Oxford University Press.

Du Boulay, Juliet. 2009. *Cosmos, Life, and Liturgy in a Greek Orthodox Village*. Limni, Greece: Denise Harvey.

Eller, Jack David. 2022. "Space Colonization and Exonationalism: On the Future of Humanity and Anthropology." *Humans* 2, no. 3: 148–160.

Fabian, Johannes. 1983. *Time and the Other: How Anthropology Makes Its Object*. New York: Columbia University Press.

Frazer, James. 1890. *The Golden Bough*. New York: Macmillan.

Geertz, Clifford. 1988. *Works and Lives: the Anthropologist as Author*. Cambridge, UK: Polity.

Geppert, Alexander (ed.). 2018a. *Imagining Outer Space: European Astroculture in the Twentieth Century*. London: Palgrave.

Geppert, Alexander (ed.). 2018b. *Limiting Outer Space: Astroculture after Apollo*. London: Palgrave.

Geppert, Alexander, Daniel Brandau, and Tilmann Siebeneichner (eds). 2021. *Militarizing Outer Space: Astroculture, Dystopia and the Cold War*. London: Palgrave.

Gorman, Alice. 2023. "Space and Time through Material Culture: An Account of Space Anthropology." In *The Routledge Handbook of Social Studies of Outer Space*, edited by Juan Francisco Salazar and Alice Gorman, 44–56. New York: Routledge.

Harvey, Brian. 2019. *China in Space: The Great Leap Forward*. Cham: Springer.

Hastrup, Kirsten and Martin Skrydstrup (eds). 2013. *The Social Life of Climate Change Models. Anticipating Nature*. London and New York: Routledge.

Helmreich, Stefan. 2012. "Extraterrestrial Relativism." *Anthropological Quarterly* 85, no. 4: 1125–1139.

Horst, Heather A. and Daniel Miller (eds). 2012. *Digital Anthropology*. London: Berg.

Ingold, Tim (ed.). 2011. *Redrawing Anthropology: Materials, Movements, Lines*. New York and London: Routledge.

Ingold, Tim. 2022. *Being Alive: Essays on Movement, Knowledge, and Description*. New York and London: Routledge.

Jeevendrampillai, David, and Aaron Parkhurst. 2021. "Making A Martian Home: Finding Humans On Mars Through Utopian Architecture." Home Cultures 18, no. 1: 25–46.

Jeevendrampillai, David, Victor Buchli, Aaron Parkhurst, Adryon Kozel, Giles Bunch, Jenia Gorbanenko, and Makar Tereshin. 2023. "An Ethnography of an Extraterrestrial Society: The International Space Station." In *The Routledge Handbook of Social Studies of Outer Space*, edited by Juan Francisco Salazar and Alice Gorman, 413–426. New York: Routledge.

Maher, Neil M. 2018. *Apollo in the Age of Aquarius*. Cambridge, MA: Harvard University Press.
Maruyama, Magorah and Arthur Harkins (eds). 1975. *Cultures Beyond Earth: The Role of Anthropology in Outer Space*. New York: Vintage Books.
Maurer, Bill. 2005. "Introduction to 'Ethnographic Emergences'." *American Anthropologist* 107, no. 1: 1–4.
Mead, Margaret. 1953. *Cultural Patterns and Technical Change*. New York: Columbia University Press.
Mead, Margaret. 1956. *New Lives for Old*. New York: Morrow.
Mead, Margaret. 1958. "Preface to Special Issue on Man in Space." *Annals of the New York Academy of Sciences* 72:167–69.
Mead, Margaret, Donald N. Michael, Harold D. Lasswell, and Lawrence K. Frank. 1958. "Man in Space: A Tool and Program for the Study of Social Change." *Annals of the New York Academy of Sciences* 72: 165–214.
Messeri, Lisa. 2016. *Placing Outer Space: An Earthly Ethnography of Outer Worlds*. Durham, NC and London: Duke University Press.
Mirmalek, Zara. 2020. *Making Time on Mars*. Cambridge, MA: MIT Press.
Mitchell, Sean T. 2017. *Constellations of Inequality: Space, Race, and Utopia in Brazil*. Chicago, IL: University of Chicago Press.
Munn, Nancy. 1986. *The Fame of Gawa: A Symbolic Study of Value Transformation in a Massim (Papua New Guinea) Society*. Cambridge: Cambridge University Press.
Munn, Nancy. 1992. "The Cultural Anthropology of Time: A Critical Essay." *Annual Review of Anthropology* 21, no. 1: 93–123.
Olson, Valerie and Lisa Messeri. 2015. "Beyond the Anthropocene: Un-Earthing an Epoch." *Environment and Society: Advances in Research* 6: 28–47.
Olson, Valerie. 2018. *Into the Extreme: US Environmental Systems and Politics Beyond Earth*. Minneapolis: University of Minnesota Press.
Olson, Valerie. 2023. "Refielding in More-Than-Terran Spaces." In *The Routledge Handbook of Social Studies of Outer Space*, edited by Juan Francisco Salazar and Alice Gorman, 31–43. New York: Routledge.
Ormrod, James S. and Peter Dickens (eds). 2016. *The Palgrave Handbook of Society, Culture and Outer Space*. New York: Palgrave Macmillan.
Parkhurst, Aaron, and David Jeevendrampillai. 2020. "Towards an Anthropology of Gravity: Emotion and Embodiment in Microgravity Environments." *Emotion, Space and Society* 35 : 100680.
Patarin-Jossec, Julie. 2021. *Manufacturing the Astronaut: An Earthly Ethnography of the International Space Station* (in FR). Paris: Editions Petra.
Petrou, Perig and Istvan Praet (eds). Forthcoming. *Social Sciences Off Earth*. Oxford: Oxford University Press.
Pink, Sarah and Juan Francisco Salazar. 2017. "Anthropologies and Futures: Setting the Agenda." In *Anthropologies and Futures: Researching Emerging and Uncertain Worlds*, edited by Juan Francisco Salazar, Sarah Pink, Andrew Irving, and Johannes Sjöberg, 3–22. London: Bloomsbury Academic.
Pozdeyeva, Irina. 1994. "A Long-Forgotten Atlantis That Never Disappeared." *Russia*, pilot ssue, 9 November 1994.

Price, David. 2020. "'Project Man in Space': Applied Anthropology's Cold War Space Oddity." *Journal of Anthropological Research* 76(3): 326–346.
Privalov, Roman. 2023. "After Space Utopia: Post-Soviet Russia and futures in space." PhD diss., Södertörn University. www.diva-portal.org/smash/get/diva2:1750044/FULLTEXT01.pdf.
Redfield, Peter. 2000. *Space in the Tropics: From Convicts to Rockets in French Guiana*. Berkeley: University of California Press.
Reid, Lauren. 2023. "Divergent Searches for Extraterrestrial Worlds on Two Mountaintops in Thailand." In *The Routledge Handbook of Social Studies of Outer Space*, edited by Juan Francisco Salazar and Alice Gorman. London: Routledge.
Riviere, Peter. 1995. "Houses, Places and People: Community and Continuity in Guiana." In *About the House: Levi-Strauss and Beyond*, edited by Janet Carsten and Stephen Hugh-Jones, 189–205. Cambridge: Cambridge University Press.
Robbins, Joel. 2013. "Beyond the Suffering Subject: Toward an Anthropology of the Good." *Journal of the Royal Anthropological Institute* 19, no. 3: 447–462.
Rosenberg, Daniel and Susan Harding (eds). 2005. *Histories of the Future*. Durham, NC: Duke University Press.
Salazar, Juan Francisco and Alice Gorman (eds). 2023. *The Routledge Handbook of Social Studies of Outer Space*. New York: Routledge.
Salazar, Juan Francisco, Sarah Pink, Andrew Irving, and Johannes Sjöberg (eds). 2017. *Anthropologies and Futures: Researching Emerging and Uncertain Worlds*. London: Bloomsbury Academic.
Scott-Heron, Gil. 1970. "Whitey on the Moon," track on the album *Small Talk at 125th and Lenox*.
Siddiqi, Asif. Forthcoming. *Departure Gates: Postcolonial Histories of Space on Earth*. Cambridge, MA: MIT Press.
Singh, Gurbir. 2017. *The Indian Space Programme: India's Incredible Journey from the Third World Towards the First*. AstrotalkUK Publications.
Space Foundation. 2023. The Space Report 2023 Q2. Accessed 9 January 2023. www.spacefoundation.org/2023/07/25/the-space-report-2023-q2.
Stewart, Kathleen. 2011. "Atmospheric Attunements". *Environment and Planning D: Society and Space* 29, no. 3: 445–453.
Stuster, Jack. 1986. *Space Station Habitability Recommendations Based on A Systematic Comparative Analysis of Analogous Conditions*. Ames Research Center, CA: NASA Contractor Report 3943.
Stuster, Jack. 2010. *Behavioral Issues Associated with Long-Duration Space Expeditions: Review and Analysis of Astronaut Journals Experiment 01-E104 (Journals): Final Report*. Johnson Space Center, Houston: National Aeronautics and Space Administration.
Stuster, Jack. 2016. *Behavioral Issues Associated With Long Duration Space Expeditions: Review and Analysis of Astronaut Journals Experiment 01-E104 (Journals) Phase 2 Final Report*. Johnson Space Center, Houston: National Aeronautics and Space Administration.
Sun Ra. 1974. *Space is the Place* [Motion picture]. 85 minutes. Plexifilm.

Szolucha, Anna, Peter Timko, Chakad Ojani, and Karlijn Korpershoek. 2023. "Ethnographic Research of Outer Space: Challenges and Opportunities." *Ethnography* https://doi.org/10.1177/14661381231220273.

Tabas, Brad. 2022. "Incipit Gaia the Grotesque Garlanded with Satellites: Re-Imagining the Anthropocene as a Post-Planetary Age." *Cosmos and History: the Journal of Natural and Social Philosophy* 18, no. 1: 131–149.

Timko, Peter, Kariljn Korpershoek, Chakad Ojani, and Anna Szolucha. 2022. "Exploring the Extraplanetary: Social Studies of Outer Space." *Anthropology Today* 38, no. 3: 9–12.

Valentine, David. 2017. "Gravity Fixes: Habituating to the Human on Mars and Island Three." *HAU Journal of Ethnographic Theory* 7, no. 3: 185–209.

Valentine, David and Amelia Hassoun. 2019. "Uncommon Futures." *Annual Review of Anthropology* 48, no. 1: 243–260.

Valentine, David, Valerie Olson, and Debbora Battaglia. 2012. "Extreme: Limits and Horizons in the Once and Future Cosmos." *Anthropological Quarterly* 85, no. 4: 1007–1026.

Vertesi, Janet. 2015. *Seeing like a Rover: Images and Interaction on the Mars Exploration Rover Mission.* Chicago, IL: Chicago University Press.

Vertesi, Janet. 2020. *Shaping Science: Organizations, Decisions, and Culture on NASA's Teams.* Chicago, IL: Chicago University Press.

Walsh, Justin, and Alice Gorman. 2021. "A Method for Space Archaeology Research: The International Space Station Archaeological Project." *Antiquity* 95, no. 383: 1331–43.

Zabusky, Stacia. 1995. *Launching Europe: An Ethnography of European Cooperation in Space Science.* Princeton, NJ: Princeton University Press.

2 Transcendence, bodies, and estranged labour in outer space

The astronaut's contribution to a general theory of hierarchy

Giles Bunch

In the early weeks of my PhD fieldwork in Germany, whenever I first told friends, flatmates, or new acquaintances that I was doing social science research at the European Astronaut Centre (EAC), they often expressed excited surprise: 'Have you met any astronauts yet?' or 'What's it like working with astronauts?' Or even 'So when are you going into space?' I would patiently explain that the focus of my research more closely follows the work of the people supporting human spaceflight activities from the ground, in particular the flight controllers and instructors working on the European contribution to the International Space Station (ISS). It would be easy to overlook these kinds of reactions and questions as simply an expression of excited intrigue around the human and technological feats of spaceflight, aligning with the common perception of spacefarers as heroic characters, embarking on dangerous missions to the 'final frontier.' But I think there is more here that should be explored. Whilst the astronauts and cosmonauts[1] going to EAC for training, press events, meetings, and other duties are clearly a crucial element of the ISS project, when I described my work, I would stress that I was interested in uncovering the side to human spaceflight that often receives less popular attention than the spacefarers inhabiting the ISS, but which is nonetheless also essential. Despite this, the excited intrigue around 'working with astronauts' often persisted.

As the research progressed, I reflected on these early conversations in relation to what I was seeing at EAC itself. I became aware of the position that the astronaut holds, as evidenced through the ways in which, on a basic level, these spacefarers are frequently spoken about, and acted towards, as though they are fundamentally special, transcendent kinds of people. For my informants going through flight controller training, seeing an astronaut in the corridor, or even better, having the opportunity to exchange a few words with one, was a noteworthy event, especially for those recruits who were new to the world of EAC. When I encountered

DOI: 10.4324/9781003437956-2

occurrences such as this, I was left with the lingering question of how it comes to be seen as so ordinary – so *natural* – to treat certain people as though they are fundamentally special. I was reminded of the opening paragraph to the first essay in David Graeber's *Possibilities: Essays on hierarchy, rebellion, and desire*:

> This is an essay about the nature of hierarchy. In it, I want to delve into hierarchy's most elementary forms: the way people avert their eyes or stand at attention, the sort of topics they avoid in formal conversation, *what it means to treat another human being as somehow abstract, sacred, transcendent, set apart from the endless entanglements and sheer physical messiness of ordinary physical existence*—and why something like that always seems to happen when some people claim to be inherently superior to others. It seems to me an investigation like this is important since it is only by beginning to ask such questions that we can begin to think about which of the qualities we ordinarily lump together in a word like "hierarchy" are really inevitable features of human social life, and which might prove dispensable.
>
> (2007, 13 emphasis added)

Following the investigation proposed here by Graeber, this chapter is an engagement with the phenomenon which sees astronauts frequently being treated as 'abstract, sacred, transcendent' beings. I aim to further our understanding of the ways in which such hierarchical social formations frequently come to be viewed as *natural*, rather than as being socially and historically constituted. Additionally, this piece is reflexively methodological in that the same anthropological questions being provoked by the status of the astronaut also affect the very methods with which social scientists investigate their subjects, both in terms of access to fieldsite data, and in the conceptual frameworks used to develop specific research questions. The main analytical component of this chapter deals with the questions: how are characteristics of hierarchy manifested in relation to the astronaut? And, what are these characteristics doing? And the reflexive component asks: what do these insights contribute to methodological approaches in the study of spaceflight contexts, and beyond?

The approach in tackling these questions takes the arguments from Graeber's essay as a starting point; in particular through considering how the body and its substances, the boundaries between persons, and the degrees to which people and labour become reified, speak to the ways that hierarchies are constituted and felt. In the opening essay of *Possibilities*; 'Manners, deference, and private property: Or, elements for a general theory of hierarchy' (referred to as *elements* from here on), Graeber revisits what he describes as the 'hoary' anthropological categories of 'joking relations' and 'relations of avoidance,' linking these to a dyadic conception of persons; as either 'a collection of bodily substances ultimately continuous with the world surrounding it, or as an abstract set of properties set apart from that world' (2007, 30–31). For Graeber's development of a theory of hierarchy, these conceptions of the person are brought into dialogue with Norbert Elias' work on the history of manners (1978), thereby linking this dyad to relative conditions of egalitarianism or hierarchy. Joking relations generally denote a more-or-less egalitarian relationship, demonstrated in the reciprocity of 'licence between joking partners' (Graeber 2007, 19). Relations of avoidance, on the other hand, are characterised as an inversion of joking, with the body being:

> ... closed, all orifices shut off and nullified; nothing flows either in or out. The body is constituted as a perfect, abstract, and self-sufficient thing unto itself, with no need for exchange either with other bodies, or the world.
>
> (2007, 19–20)

In this chapter, I won't be using the dyadic model from *elements* as a rigid guiding schematic, but instead its use applies more generally to the characteristics of astronauts, and to the nature of hierarchical relations in this context. To this end, we begin with an exploration of the complexities associated with the body and the person of the astronaut; of the ways in which they inhabit a dual positionality (Olson 2018, 87) as both sacrificial test subject and elite individual. This is developed through an analysis of the sovereign qualities that such a positionality gives rise to, revealing how the elite status of the astronaut delimits certain freedoms. This is followed by an examination of the working relationship between ISS crews and the flight control team, and the qualities of abstract, fungible labour which characterise the astronaut's work when on the station. Through this approach, which frames such activity as estranged labour, the interplay between these qualities of work and the stated metavalues of human spaceflight further reveal what hierarchy might be doing in this context as well as in other scenarios. The methodological reflections – applicable to spaceflight contexts and beyond – follow after the ethnographic analysis. Structuring the chapter in this unorthodox way allows for these contributions to respond to the earlier anthropological arguments.

Before moving on, it should be noted that, unless stated, this chapter focuses on NASA and ESA astronauts and activities taking place in the United States Orbital Segment (USOS)[2] of the ISS. For readers interested in ethnographies that include cosmonauts, I'd recommend recent work by Julie Patarin-Jossec (2020; 2021; 2022), as well as Jenia Gorbanenko's upcoming study as part of the ETHNO-ISS project at UCL (Jeevendrampillai et al. 2023, 420).

Astronaut bodies, experimental subjects, and propriety

The body is centrally important for thinking about how characteristics associated with hierarchy become manifested in the seemingly oppositional qualities of the astronaut. In popular discourse, astronauts are commonly viewed as exemplifying the best qualities of humanity, chosen representatives exhibiting the apex of physical and intellectual ability (Weibel 2024), and 'envoys of mankind' (Dordain et al. 2004, 10) in the future-oriented project of exploring worlds beyond our own. Bundled up in this is the astronaut's role as the biomedical test-subject, whose body serves the production of data-points associated with the impacts of long-duration[3] spaceflight on the elite human (Weibel 2019). At first gloss, they abound with contradictory qualities: both transcendent being and sacrificial subject, their person being closed off, contained, subject to stringent privacy regulations, whilst their bodily substances are given out or extracted in the name of research. Astronauts attract popular questions about their bodily substances in space; almost obsessive queries, for example, on how they urinate, defecate, or menstruate, questions that would be unthinkable to ask in many other contexts. As Valerie Olson states, the astronaut inhabits 'a biopolitical limbo somewhere between privilege and abjectness' (2018, 87). This is a very apt description, to which I would add that rather than being *between* privilege and abjectness, they perhaps are both *at the same time*.

Here, I introduce the term 'bodily sovereignty' when discussing the astronaut's negotiation of how their body is used and scrutinised. This use of 'sovereign' deliberately follows Agamben's discourse in *Homo Sacer*, by which sovereign power is located in a ruler's or governing body's capacity to indefinitely suspend laws (1998, 15). Being 'sovereign', therefore, is to possess a power of exception; it is the status of being both bound by a set of laws, as well as having the freedom to decide when and how those very laws are suspended. It is a condition of being both 'outside and inside the juridical order' (1998). In my usage, I extend Agamben's definition with 'bodily sovereignty' to describe the condition by which an astronaut might hold the power of exception over their own body, as well as having the capacity to decide what happens to it. Where my use of sovereignty in this piece differs from Agamben's description however, is that whilst being sovereign in a legal sense is to possess a monopoly over the power of decision, the 'bodily sovereignty' of the astronaut is instead a contingent and compromised state of being. Bodily sovereignty in my usage is not a total power of exception, but rather a measure of value that is constantly challenged and contested. Simply put, bodily sovereignty is not something that astronauts possess in abundance.

At EAC, the main European Space Agency (ESA) training centre for European activities linked to the ISS, I observed how the bodily sovereignty of astronauts is configured. The impacts of living in Low Earth Orbit (LEO from here on), and the effects of prolonged exposure to microgravity on the body, are a valuable source of data for the space agencies involved in the ISS. During fieldwork, some of my informants were involved in 'Direct Return,' the process by which the gathering of such data continues after ESA astronauts return to Earth. Following 'splashdown' near the Florida coast, the astronaut's work is far from finished, and they return to EAC and the neighbouring space medicine centres in Cologne, Germany, to begin the vital process of gathering human physiological data within specific time-critical windows. This is busy, strictly managed, time-sensitive work. Astronauts have their blood drawn, urine samples taken, and various other physiological variables measured, alongside recuperative activities designed to aid recovery after long-term space living.

The example from Direct Return raises the question: what is an astronaut's capacity to refuse procedures? Continuing with 'bodily sovereignty,' I'd further characterise this term as denoting an astronaut's *potential* to refuse. The emphasis on potential is important here because it describes exactly that: the *potential* for something which rarely *actually is*. Direct Return offers an example of such, since it was rare, but not unknown for ESA astronauts to renegotiate the schedule of tests and recovery activities by asking to re-arrange the timetable and carve out some time off. Such instances are illustrative of a tension between the priorities of the biomedical researchers who are eager to obtain their data in the correct time-specific windows, and the astronauts who seek some time away from the continuing, highly managed schedule. NASA astronauts have similarly been known to resist intrusive medical procedures, in some cases being allowed to refuse monitoring that might reveal conditions leading to them being disqualified from flight (Olson 2018, 87). The *potential* of refusal in relation to the bodily sovereignty of the astronaut also finds expression in the relationship between the flight control team (FCT from here on) and astronaut crews on the ISS. Specific polite forms of speech echo the logic of 'relations of avoidance' as described in *elements*, the indirect qualities of which invite such a *potential* of refusal. To this we turn next.

The ways that Eurocom (a portmanteau of European Communicator) operators in the FCT are trained during simulations speaks further to how astronaut sovereignty is configured, as well as to hierarchy's 'most elementary forms,' to use Graeber's terminology (2007, 13). Before we get into this though, some exposition of the details is required for clarity. 'Eurosims' are a key component of the training for ISS flight controllers on the European side of the FCT. They simulate a day of 'operations' as they would take place (for want of a better expression) in 'real life'[4] on the ISS. Trainee flight controllers, in their respective positions and working in control centres across Europe, coordinate the running of simulated, scheduled experiments and maintenance activities. And whilst trainees in the Eurosims use tools and interfaces in much the same way that they would in 'live' operations, the astronauts in these simulations are instead mostly performed by instructors and referred to as 'surrogate' astronauts. Instead of being on the ISS itself, these surrogates often sit at one of the workstations in the simulation room; so, when a surrogate is 'working on an experiment in the *Columbus* module' as part of the role play, in reality they might be in the same room as the trainees, sitting a few metres away from them. In the following account from the Eurosim, the reader should keep in mind this 'role-playing' aspect. For example, when I describe 'an experiment involving an astronaut in the *Columbus* module,' this is a description of events as they occur within the *narrative logics of the simulation.*

The episode in question occurred during a Eurosim and involved an experiment with a piece of wearable technology used for gathering physiological data from an astronaut. According to the procedure, the astronaut had to remain in the *Columbus* module for ten consecutive minutes, otherwise the equipment would stop gathering data. As a test for the trainee flight controller responsible for the experiment, the astronaut decided to leave *Columbus*, resulting in an interruption to the experiment. It was agreed between the *Columbus* Flight Director and the trainee responsible for the activity, that the experiment would be run again. Meanwhile, the trainee Eurocom was following this exchange over the 'loops' (the voice channels that are used between operators in the FCT to coordinate between themselves), in anticipation of informing the astronaut that they would need to perform the experiment again. The Eurocom position serves as the single channel of verbal communication over the 'space-to-ground loops,' between astronauts on the ISS and the European FCT, and so their training covers these associated practices. Before the Eurocom was given the go-ahead to inform the astronaut (reminding them to stay in *Columbus* during the ten minutes), they were asked by the *Columbus* Flight Director to rephrase the directive. Rather than a more direct 'we need you to…' or 'can you…,' for example, it was recommended that they should couch it instead as a polite *request* over the space-to-ground loop. What might at first appear as a minor detail, I saw expressed as a pattern, and this trend of couching instructions as requests emerged in the use of phrasings such as 'at your convenience…' when instructing an astronaut crew member to perform a necessary task.

What is significant about these exchanges is the fact that the astronaut would never normally refuse such an instruction, and yet the suggested re-phrasing of instructions as requests opens up – within the logic of the wording – the possibility of refusal. The exchanges on the loops within the team (to which astronauts are not privy) before the request was made by the Eurocom, were couched in a language that assumed the crew would carry out the task; put bluntly, in this register, the astronaut is a resource to be instructed. The recommendation from the Flight Director to the Eurocom to rephrase the directive tacitly cites the need for the performance of 'emotional labour' (Hochschild 2012) on the part of the communicator, shifting the register to that of a request.[5] Both Eurocoms and astronauts have commented on the dynamic formed by these speech patterns. From their conversations with crew members on the ground, Eurocoms are aware that astronauts dislike the excessive *politesse*[6] in these communications (see also Stuster 2010, 2016 for commentary on 'praise inflation'), wishing instead 'to be treated as normal people' (Interview 2022). Eurocoms have also expressed frustration at the amount of 'back-and-forth' that sometimes goes on over the loops when deciding on the wording used to *request* an astronaut to perform a simple task on the ISS.

I suggest that the two registers by which flight controllers on the one hand *speak about*, and on the other hand *speak to* the astronaut, are expressions of a relation of avoidance, as described in *elements* (Graeber 2007, 19–20). Discussions over the loops between the FCT frames the ISS crewmember as abstract labour (discussed further in the next section), through language which indicates an assumption that the astronaut will carry out the task. The subsequent use of indirect speech ('at your convenience…') in the exchanges between Eurocoms and ISS crewmembers, reinforces the logic of avoidance since both the astronauts and the flight controllers know that what is being asked is an entirely reasonable and ordinary request in the context of ISS operations. Here, a comparison with a case study in which these logics are far more pronounced can help to illustrate more clearly how the *politesse* on the side of the Eurocom aligns with it being understood as a relation of avoidance, and as indicating a hierarchical relationship.

In *The Pot-King*, Jean-Pierre Warnier (2007) describes the ways that the king of Mankon is fenced around with general prohibitions relating to his body: context defined rules relating to who is allowed to touch, speak to, or share food with him, for example. Prohibitions on speaking to the king can be provisionally suspended, but only when specific practices are followed:

> …one has to screen off one's breath and speech by putting one or both hands in front of one's mouth or else, one's clenched fist. The king, by contrast, speaks to his subjects in such a way that his breath and speech reaches them directly.
>
> (Warnier 2007, 172)

In this example from Warnier in the western highlands of Cameroon, the relations of avoidance as indicators of a hierarchical relationship are particularly pronounced when compared with my own fieldsite, but nonetheless share a similar logic. The *politesse* in the wordings (the indirect speech) that the Eurocom is encouraged to use over the space-to-ground loops, echoes the screening off of one's breath (an embodied act of making the voice indirect) when speaking to the king. The second observation is that the king of Mankon is the one who – as per the nature of being *sovereign* – does not need to observe those prohibitions, and as the superior in the hierarchy, has the licence to suspend them. I'd argue that when astronauts comment on the excessive *politesse* of vocal communications, wishing to be 'treated as normal people,' they do so from a position that grants them that licence. If the bodily sovereignty of the astronaut is a measure of the power of exception, then it is precisely their elevated status that allows them to express the desire to be acted towards in a different way.

Further to the themes of monarchs and sovereignty, these can additionally tell us about the ways that hierarchies coincide with certain limits on the freedoms of abstract, transcendent beings. As noted earlier, resistance to the timetabled demands of tests and recuperative activities during Direct Return is unusual, and the re-phrasing of directions as requests during space-to-ground communications invites the possibility of refusal, without such a refusal ever realistically becoming enacted. British constitutional monarchy offers a helpful point of comparison here. Political scholars Robert Hazell and Bob Morris describe the distinction between 'prerogative powers' held by government ministers, and those powers still remaining in the hands of the Queen[7] (2017). What is peculiar about these 'reserve powers' is that whilst they are formally and *potentially* substantive – granting the monarch the ability to appoint and dismiss ministers, and summon, prorogue and dissolve Parliament, for example – the *actual* capacity to act on these is so limited to extremely contingent circumstances in the contemporary era, that it renders them almost non-existent (2017, 7; see also Sahlins and Graeber 2017, 8). The example from British royalty helps us consider more generally how forms of unique status – of being somehow elevated, transcendent, or otherwise set apart – is also to have one's freedoms highly bounded, again inviting a return to Agamben and his arguments around the nature of becoming sacred, outlined below (1998).

Here, it is necessary to address the notion of sacrifice as it has been discussed by others in relation to the astronaut. The first version of this, proffered by anthropologist Deana Weibel, is of sacrifice being enacted by the ground teams, mission support specialists and other spaceflight workers, which enable the astronaut to perform their work in space. In her words, it is 'an example of a cultural phenomenon where a large group of people mobilizes and sacrifices to allow a smaller group to achieve something unavailable to the population at large' (Weibel 2019). The second version, and the one with which this chapter is more concerned, comes from the application of anthropological insight directly to human spaceflight in the European context. In her contribution to the 2004 symposium 'Legal and Ethical Framework for Astronauts in Space Sojourns,' Monika Konrad describes the astronauts themselves as being sacrificed in the service of biomedical research:

> Many astronauts sacrifice their time, privacy, and indeed their entire physical selves to space science: constantly giving blood, wearing sensors, logging their food and drink, storing their excrement for study and submitting themselves and each other to all kinds of indignities, and perhaps dangers.
>
> (Dordain et al. 2004, 32)

The activities described here by Konrad relate to the gathering of physiological data under conditions which, by several measures, are harmful to human health, hence inviting the interpretation of these being acts of personal sacrifice. My modification to this would be that given the ways that astronaut health is configured in the field of space medicine, it becomes difficult to retain the 'sacrifice' descriptor used by Konrad. Olson's work carried out during the early years of the ISS draws attention to the ways that space medicine problematises both human life and environments to reconfigure what is defined as 'healthy' for the astronaut (2010). This is described by her informants as 'space normal,' a dynamic state in which astronaut health is determined by 'human/environment interactions that are in constant flux and cannot be understood as "natural"' (2010, 182). Given the damaging effects of long-duration spaceflight on astronaut health (Krittanawong et al. 2023), 'space normal' serves as a means for redefining the boundaries of acceptable exposure to the detrimental impacts of outer space. In short, as an aerospace physiologist explained to Olson: 'We need to define "space normal" so we don't keep trying to treat astronauts in space as if they're sick' (2010, 170).

Returning to Agamben, 'space normal' operates in a similar logic to the philosopher's discourse around the nature of sacred life, as being 'life that cannot be sacrificed and yet may be killed' (1998, 82). The process of redefining the boundaries of acceptability by which an astronaut on the ISS is permitted to receive harm due to long-duration spaceflight, also by extension excludes the framing of astronauts as performing personal sacrifice. This gets us closer to an understanding of how the astronaut can be understood as being sacred. They have been excluded from categorisations that would consider them as receiving harm, as being treated 'as if they're sick' through their very consecration *as astronauts*. A rephrasing of Agamben for this context might read something like this: the sacred astronaut cannot be sacrificed and yet is sanctioned to receive harm from the very conditions of their *being an astronaut*.

To conclude this section, we have started from the position that astronauts are acted towards and spoken about in a way which positions them as transcendent or sacred beings. Their work on the ISS itself, and during the months post-flight, is punctuated by what Konrad has described as the personal sacrifice of having one's body serve the gathering of physiological data around the impacts of long-duration spaceflight. To make an obvious point about astronauts that I've not yet openly stated in this chapter: the ways in which they are *set-apart* are literally prefigured by their being the only people in the nexus (Buchli 2021) of teams and personnel linked to the ISS that leave Earth's surface. As a consequence of the ways that human health is configured for long-duration spaceflight, this nature of being set-apart is further compounded by their exclusion – via a reading of Agamben – from being sacrificed, but instead having their harm sanctioned. The verbal patterns observed over the space-to-ground loops highlight the physical distance between operators on the ground and astronauts on the ISS, as well as indicating the relation of avoidance emerging out of astronauts' sacred status. The relationship between Eurocom and those working in orbit is also indexical of a wider consideration of the astronaut's relationship to their labour on the ISS. To this we turn next.

Estranged labour on the ISS

> ... you need to be able to be OK with the fact that you are working all day from a timeline, from instructions that you're given. There is very little room for, you know, personal intervention or creative inputs ... or even of understanding; ninety percent of the things that I do every day, I don't know exactly what I'm doing, and I don't know why I'm doing it. I just know that this is what I have to do, and I do it. And you have to be OK with that. Some people would find that very frustrating.
>
> (Interview with an active astronaut, 2022)

> A procedure is basically a work instruction for the astronaut to perform a certain activity. It's very structured, it tells them exactly what to do. If you switch off your brain and just do what is in the procedure, you'll get the right result. Thinking too much, sometimes, is detrimental to the result because then you try to do shortcuts or things like that.
>
> (Interview with a member of the instructor team at EAC, 2021)

This section engages with the qualities and organisation of work taking place on the ISS, its relation to the complexity of astronaut positionality, and how these characteristics connect to hierarchy. The argument looks at how activities aboard the ISS can be considered forms of estranged labour in which the astronaut is largely interchangeable in terms of their 'generic skills' training, as well as in the ways that tasks are heavily proceduralised and rigorously scheduled. The discussion is brought into dialogue with the Marxian concept of *species-being* (Marx 1992, 329) as a means for exploring what the wider implications of the organisation of labour might have to say to the purpose and promises – what are described in this section as 'metavalues' (Graeber 2013, 233) – of human spaceflight more generally.

The quote from the astronaut above helps our understanding of what working in the USOS of the ISS is like. Expanding on a few of the points in the quote, the 'timeline' refers to a display showing the activities for each ISS crewmember, along with related information about operations on the ground. It is viewed on laptops by the crew, and is one of the most important windows displayed on a flight controller's screen when they are 'on console.' During observations of live shifts and simulations in control rooms at EAC, I saw the interminable movement across the timeline of the 'red line' (Contella et al. 2018, 11), a vertical indication of the time for 'now' (synchronised to Greenwich Mean Time). As this moved right across the display, coloured blocks, each representing an activity, changed to grey, indicating that a given activity had been marked 'complete.' Occasionally, some blocks turned grey before being reached by the red line if a crewmember had got to the work early, or a block remained green (indicating it as 'active') after the line had passed, if the activity was still being worked, for example. The NASA publication *The International Space Station: Operating an Outpost in the New Frontier*, written and edited by US members of the flight control team (Contella et al. 2018), includes short sections written by USOS astronauts about their experiences. Sunita Williams describes the feel of 'chasing the red line,' in which a day's activities aboard the Station are punctuated by the awareness that the line is 'either chasing you or you are chasing it' (Contella et al. 2018, 29, 374). Statements like these, when paired with more candid anonymised statements from ISS crews in the *Astronaut Journals Experiment* (Stuster 2016, 24–25), indicate the extent to which the ever-rightward creep of the red line denotes labour on the ISS as being strictly dictated by 'clock time' (Thompson 1967; Postone 1996; Snyder 2019), with one respondent to the *Journals* project explicitly stating: 'I know we're not supposed to care, but it is very difficult to ignore that clock' (Stuster 2016, 24) (see Figure 2.1).

Considering how work is managed, how activities are performed according to the ever-progressing movement of the timeline, and the ways

Figure 2.1 A workstation used by a Eurocom trainee during a Eurosim. Note the timeline viewer on the right-hand side of the screen on the left. Photograph by the author.

that knowledge and expertise of tasks is concentrated in specialist teams on the ground, spaceflight work on the ISS, I argue, is an almost archetypal expression of Frederick Taylor's proposals (originally published in 1911) for the reorganisation of industrialised labour processes according to principles of 'Scientific Management' (Taylor 1967 [1911]). Taylor's model, which found later expression in the development of Fordism, sought a reorganisation of labour based on an explicit de-skilling and capture of workers' intimate knowledge of production processes; tasks were measured, standardised, and proceduralised according to the clock-time of a stopwatch, with this knowledge being placed into the hands of an expanded management layer within the organisational structure of the factory. The reorganisation of industrialised production developed in the decades following the First World War, with mid-twentieth-century commentators describing its emergence as a 'social system' in which the 'advanced planning for production, operating and control' also saw the increasing subdivision of labour and the professionalisation of a management class (James, Dunayevskaya, and Boggs 2013).

Returning to the interview quoted above, the similarities between Taylorism, and what the astronaut is describing is striking: the assessment that their everyday work on the Station is almost entirely proceduralised, strictly timelined, and performed without the requirement of much specialist scientific knowledge for each experiment, aligns closely with the tenets of Scientific Management. And although examples of discrete artisanal skills have been documented in the production of artefacts on the ISS (see Victor Buchli's forthcoming work with the ETHNO-ISS project: Jeevendrampillai et al. 2023, 415), the overarching logics still remain. There is not the scope here to fully explore the organisational history of how activities are arranged on the ISS, but what can be said is that the current arrangement follows a clearly defined division of labour between the ground teams and the crew. Ground teams plan activities, prepare operational documents, activate equipment and systems remotely, and observe crewmembers, providing directions and advice during procedures where needed (think back to the Eurocom from the previous section). The above quote from the instructor team member, in which the qualities of ISS procedures are described as precise, structured and unambiguous, further supports a view of how we can understand such a division of labour. Research teams and their Principal Investigators develop experiments or what are known as 'technology demonstrators,' with the directions for how these are operated on-orbit being laid out in a procedure in the most precise, unambiguous way possible. In terms of their basic training, rather than being instructed in the specific techniques required of each and every activity an astronaut might perform on the ISS, they are instead trained according to the requirements for 'generic skills' in a training regime which seeks to standardise – to *level* – the skills between them.

In framing astronaut work as estranged labour, I am drawing explicitly upon Marx's earlier work and his arguments on estrangement found in the *Economic and Philosophic Manuscripts of 1844* (referred to as *1844 Manuscripts* from here on) (Marx 1992, 322–334). This estrangement, the sense that astronaut labour is alien to them, is foreshadowed in the opening quote: 'I don't know exactly what I'm doing, and I don't know why I'm doing it,' and is further evidenced (as described above) in the qualities of how work is performed and organised on the ISS. In the later part of his commentary on estranged labour in the *1844 Manuscripts*, Marx develops the notion of human *species-being* (Wartenberg 1982; Marx 1992, 327; Dyer-Witheford 2004). In these passages, Marx defines species-being as the state by which '[m]an makes his life activity itself an object of his will and consciousness,' as an expression of his 'conscious life activity' (Marx 1992, 328). For Marx, this is what characterises human action, in contrast with other animals – the capacity to *objectify* one's own life, to produce, through design, and then to reflect on the nature of that process, as comprising of humans' collective and individual life-processes:

> It is therefore in his fashioning of the objective that man really proves himself to be a *species-being*. Such production is his active species-life.

> Through it nature appears as *his* work and his reality. The object of labour is therefore the *objectification of the species-life of man*: for man reproduces himself not only intellectually, in his consciousness, but actively and actually, and he can therefore contemplate himself in a world he himself has created. In tearing away the object of his production from man, estranged labour therefore tears away from him his *species-life*, his true species-objectivity, and transforms his advantage over animals into the disadvantage that his inorganic body, nature, is taken from him.
>
> (Marx 1992, 329 original emphasis)

Thinking through the estrangement of astronaut labour via this engagement with the notion of species-being, provides the means by which we can now connect issues around the status of the astronaut (as a transcendent, abstract, sacred kind of being), with the role that they serve as biomedical test subjects described in the earlier section. A helpful but somewhat idiosyncratic way of drawing these connections out is through the engagement with a critique that could easily be made of my framing of astronaut work as estranged labour. This critique is that, although the everyday experience of work aboard the ISS is generally highly proceduralised and uncreative, comprising of a scientific content that is often unknown to those performing the experiments, the astronauts, however, regularly embody and perform a set of metavalues (Graeber 2013, 233) associated with the aims of ISS research and human spaceflight more generally. These metavalues are to be found in the stated benefits of carrying out science in LEO. Biomedical and biological research on the ISS (for which, as we saw, the astronaut body is essential) is frequently cited as benefiting Earth's inhabitants through the improvement of pharmaceuticals, medical procedures, or the production of prostheses, for example (Castaño 2023, 427; Jeevendrampillai et al. 2023, 416). Much of the biomedical research also serves the metavalue of continued and expanding human exploration of the cosmos, with this research providing the basis for techniques of habitation in outer space. In the contemporary moment, the astronaut is to be an 'envoy of mankind' in the further exploration of the Moon and Mars. By this measure, so the potential critique goes, the 'estranged labour' frame cannot fully fit, because these metavalues are an expression of the ways in which the organisation of micro-scale labour practices, as described above, are productive of the future-oriented project of improving life for Earth's inhabitants and exploring other worlds.

Here is where the notion of species-being intervenes. Taking Dyer-Witheford's analysis of the term, a crucial element of species-being is the realisation of the 'historical possibilities of self-development' (2004, 4) through the objectification of individual and collective enactment of human 'conscious life activity' (Marx 1992, 328). At the first analysis, the metavalues of astronaut labour (scientific benefits for Earth's inhabitants and enabling the exploration of the cosmos) would fit such an understanding of species-being because these are an articulation of a quasi-*historical* project. A project in which the stated overarching aims of human spaceflight justify and explain the everyday experience of astronauts on the ISS. My argument, however, is that the metavalues of human spaceflight aren't an articulation of the historical character of species-being, precisely because they are metavalues which struggle to express anything outside of themselves. What I mean by this is, if we consider the metavalue of astronaut work on the ISS as enabling further space exploration, especially beyond LEO, then that metavalue is speaking to its own continuation, the pursuit of *more spaceflight* (Graeber 2013, 225–226). It struggles to articulate anything outside of itself because it pursues its very own action. The metavalue of astronaut work on the ISS as providing benefits for those 'back on Earth,' I argue, operates as a second-order justification for the previous metavalue. Given the political-economic context in which these activities take place, it would be very difficult to justify the resources that go into human spaceflight for the sake of exploring the cosmos in its own right. The resourcing for such projects requires the political justification that is produced by the promise of those benefits. Furthermore, these metavalues are often expressed in a way that is universalising, non-specific, and repeatedly deferring to an as-yet-to-be-realised future (David Jeevendrampillai's forthcoming study with the ETHNO-ISS project deals excellently with these themes: Jeevendrampillai et al. 2023, 419, see also Jeevendrampillai 2017). The material benefits are promised for a singular 'humanity' on Earth, sidestepping any acknowledgement that those benefits will likely not be felt universally. These metavalues struggle to express anything outside of themselves because they rarely speak to the historical specificities of their action. If the supreme metavalue of human spaceflight is the further exploration of worlds beyond our own, then, in our current period this is ahistorical because it rarely engages with the specificities (beyond the realm of the empirical) of how such a life is to be lived. If this is true, then a disjunction can be seen between the embodied, micro-scale activities of astronauts on the ISS (what I've described as estranged labour) and the 'historical possibilities' (Dyer-Witheford 2004, 4) that might be afforded by such.

My argument is that the qualities of the astronaut as a sacred, transcendent kind of being contribute to the production of such a disjunction. This is evidenced in the details already given: as a sacred test subject, their sovereignty is highly compromised through the ways in which they give over their body and their substances for the production of biomedical data, with their symbolic stature functioning to further compound such a process of reification. The fact that astronauts occupy such a positionality, and that their labour on the ISS is estranged, produces a situation in which the universalising, non-specific qualities of the metavalues therefore become *the only thing remaining* that can be said about human spaceflight in this context, because such metavalues cannot speak to values outside of themselves. As an illustration, the popular debate surrounding the so-called 'strike' or 'mutiny' in December 1973 of the Skylab-4 crew hints at the point I'm trying to make. The accurate telling of events is not what interests me here, but rather how the story of the strike indicates a popular, persistent desire towards an interruption of the otherwise estranged activities that the crew were engaged in. This is shown in the vast divergence in the stories surrounding the mission. For NASA historians, 'no strike or mutiny took place' (Uri 2020), and whilst there was a re-negotiation of scheduled workload during the mission, the claims that the crew deliberately turned off their radio and 'downed tools,' so to speak, are most likely the product of myth. That such myth has endured, however, is an indication of an appeal found in the suggestion that, for a brief moment, the crew of a space station fully refused the conditions of their work and briefly produced a novel kind of politics that exceeded both their micro-scale estranged labour *and* the non-specific metavalues associated with human spaceflight.

By way of tying the arguments from this chapter together, we are left with three approaches for understanding the astronaut which present a means for better understanding the ways that hierarchy functions. In the first case, we can understand the astronaut as a *symbol* or a representative *figure*, of the future-oriented promises – the metavalues – of human spaceflight, expressed in the excited intrigue (described in the introduction) that I've encountered when describing my fieldsite. It's also evidenced in astronauts' numerous political and public relations appointments, in which they serve as the representatives or spokespeople for the exotic and unique qualities of human spaceflight (see, for example, Melvin 2018, 144). The second approach is that of being *sacred*, as described in the section on the body, by which astronauts' bodily sovereignty is compromised as a matter of course in the production of physiological data on the impacts of space living. Such a state of being sacred, is akin to particular characteristics of constitutional monarchy, by which the very status of being the thing that they are – that is, the status of being an astronaut – brings with it a slew of limits upon their freedom. Hence the status of their sovereignty being characterised as the *potential* of refusal. And finally, following the description of work on the ISS, we have an approach of understanding the astronaut as *abstract*, in the sense that spaceflight work is commonly estranged labour. Taken together, we begin to see how the act of treating someone as fundamentally 'abstract, sacred, transcendent,' and the practices which compound such a situation, contribute to a disjunction between the specific qualities of the labour performed on the ISS, and the historical possibilities of this activity.

On methods

The motivation for this chapter emerged out of a recognition of 'hierarchy's most elementary forms' (Graeber 2007, 13), in the ways that astronauts were often spoken about and acted towards during my fieldwork at EAC. Future studies in, and more importantly beyond, the anthropology of outer space, would do well to remain attentive to these details which are subtle or easily dismissed. General questions emerging retroactively from my own fieldwork can apply to any context: is there a person or a group of people that when they are present, attract furtive glances, attention, or comments? Are there ways that a person is somehow bounded off, or otherwise protected? When arranging to meet someone, is this process mediated by prohibition or ritual (for example, arranging the appointment through an assistant)? Are there ways that someone might stand differently or change their register of speech when interacting with a specific person? And so on. Even slight alterations to the built environment can give clues to when someone is treated as somehow *special*. At EAC, this is apparent in a block of white tape strips that have been attached (as an intentioned addition) to a window which closes off the view between the *Columbus* mock-up in the Training Hall and a large staircase in a communal area. Instructors informed me that this was added to discourage people looking in during astronaut training since it was a frequent distraction for those delivering classes. Obviously, reading these cues is the work of anthropology in that there can be multiple registers of interpretation based on the researcher's grasp of language (Astuti and Bloch 2012, 456–57) and understanding of bodily/verbal articulations (or as Warnier would call them, 'sensori-motor conducts'), which are modulated by emic and etic factors such as gender, culture, class, ethnicity, and so on. Just looking at some of the anthropological literature on emotion (to pick a few of several examples) demonstrates the extent of interpretive work involved (Wikan 1989; Parkhurst and Jeevendrampillai 2020; Wierzbicka 1986).

The cues that I've suggested in the questions above are important for detecting hierarchy, particularly in contexts where the existence of this is otherwise unacknowledged or even outright denied. There are two main challenges presented by this. The first is to the choice made by the anthropologist of whether to take such denial of hierarchy at surface value and repeat these claims, or to adopt the position of the Shakespearean fool, someone who is in the confidence of power-holders around them, but who also has licence – in its way, also a form being sovereign – to say exactly where they think power relations lie, and what the associated hierarchies are doing. My preference is for the more difficult, latter approach. The second challenge is from what Vita Peacock has described as the 'negation of hierarchy' (Peacock 2013, 2015), by which anthropologists, especially those working on organisational studies from the late twentieth century on, have in general retreated from tackling questions around social hierarchy in favour of discourse on 'action played out through the use of terms such as "agency," "action" and "practice"' (Peacock 2015, 12). One of the consequences of this trend is that it provides inadequate tools for fully interrogating how wider social forces shape, and are shaped by, the 'action' of those within an organisation, with 'any supra-personal entity' appearing as 'the secondary product of such action' (2015). The potential result being that such a 'negation of hierarchy' also curtails anthropology's capacity to detect hierarchies and speak to the ways these shape the lives of those we study.

The astronaut then, can be taken as an extreme case study in a context in which the 'elementary forms' of hierarchy are encountered. Aligning with the spirit of the proposal put forth in *elements*, I'd argue that the kinds of characteristics I've described for the astronaut most likely share several logics with other contexts in which individuals or groups come to be treated as though they are fundamentally sacred, abstract, or set-apart. The positionality of athletes within elite team sports is a particular case in point (see, for example, Canada 2022). In taking the astronaut as a case study in which the characteristics I've described above are particularly pronounced, there is another issue at stake which relates back to the metavalues of human spaceflight. If the metavalues that I've described above have universalising tendencies (that is, conducting science in orbit to *benefit inhabitants back on Earth*), then this reinforces an idea (borne out in the 'envoys of mankind' label) that astronauts are *representatives of humans as a totality*. The political implications of this are that, if astronauts are the universalised representatives of Earth's inhabitants, then the social and organisational configuration of their work that I've described, also risks becoming totally naturalised, thereby occluding a discursive space for the enactment of alternate political avenues and possibilities. As I've tried to show in the previous pages, the characteristics associated with hierarchy frequently produce alienated social relations. But the extreme qualities of the astronaut also invite the possibility of bringing into sharp relief – echoing the quote from *elements* in the introduction – what features of human social life linked to hierarchy really are inevitable, and what might be dispensed with.

Acknowledgements

This research has been produced as part of the European Research Council-funded ETHNO-ISS project based at University College London (grant number 833135). I'd like to thank the editors, Jenia, Jeeva, and Adryon for their support, encouragement, and thorough feedback throughout the many drafts of the piece. Victor Buchli gave wonderful text recommendations that helped shape the theoretical direction, and my continuing discussions with Aaron Parkhurst about astronauts were a vital source of motivation. Chloe Hough's intelligent, and ever-accurate, insights about power, hierarchy, estrangement, and labour have been a major influence on this piece. Lastly and most importantly I want to thank all the anonymous research participants at EAC, Col-CC and elsewhere, who have shown me immense generosity and support throughout my PhD research.

Notes

1 The term given to spacefarers from the Russian Federation, former Soviet Union, and Soviet-adjacent states.
2 The ISS is made up of two main segments: the Russian Orbital Segment (ROS), located aft, and the United States Orbital Segment (USOS), located forward. The USOS is comprised of modules and other infrastructure affiliated with the US, Japan, Europe, and Canada.
3 Here I use 'long-duration' to denote the typical length of missions on the ISS. Astronauts and cosmonauts commonly stay on the ISS for roughly six months at a time, this period being sometimes shorter or longer depending on a range of factors.
4 This is by way of a shorthand. There are plenty of interesting things to be said about the roleplay and fantasy elements of the Eurosims through an engagement with the likes of Bateson (2006), but there is not the room to do this here.
5 Gender is definitely an important factor in my study of EAC and Col-CC. Unfortunately, there is no room to discuss it here.
6 Here I've taken the use of 'politesse' from Victor Buchli, who used the term during ETHNO-ISS team meetings to describe similar phenomenon amongst his informants.
7 Following in the spirit of Kantorowicz (1997), it is unnecessary to make any distinction between Elizabeth II and Charles III.

References

Agamben, Giorgio. 1998. *Homo Sacer: Sovereign Power and Bare Life*. Translated by Daniel Heller-Roazen. Stanford, CA: Stanford University Press.

Astuti, Rita, and Maurice Bloch. 2012. 'Anthropologists as Cognitive Scientists'. *Topics in Cognitive Science* 4: 453–61. https://doi.org/10.1111/j.1756-8765.2012.01191.x.

Bateson, Gregory. 2006. 'A Theory of Play and Fantasy'. In *The Game Design Reader: A Rules of Play Anthology*, edited by Katie Salen and Eric Zimmerman, 314–28. Cambridge, MA: The MIT Press.

Buchli, Victor. 2021. 'Extraterrestrial Methods: Towards an Ethnography of the ISS'. In *Lineages and Advancements in Material Culture Studies: Perspectives from UCL Anthropology*, edited by Timothy Carroll, Antonia Walford, and Shireen Walton. Abingdon: Routledge.

Canada, Tracie. 2022. 'Plantation Logics at the NFL Combine'. *Anthropology News*.

Castaño, Paola. 2023. 'Plant Biologists and the International Space Station: Institutionalising a Scientific Community'. In *The Routledge Handbook of Social Studies of Outer Space*, edited by Alice Gorman and Juan Francisco Salazar, 427–40. Abingdon: Routledge.

Contella, Dina E., David H. Korth, Michael L. Lammers, Courtenay R. McMillan, Emily Nelson, Royce J. Renfrew, Brian T. Smith, Scott A. Stover, and Edward A. Van Cise. 2018. *The International Space Station: Operating an Outpost in the New Frontier*. Edited by Robert Dempsey. Houston, TX: National Aeronautics and Space Administration. www.nasa.gov/sites/default/files/atoms/files/iss-operating_an_outpost-tagged.pdf.

Dordain, Jean Jacques, Marcio Barbosa, Philippe Achilleas, Monica Konrad, Agathe Lepage, Jacques Arnould, Juan Manuel de Faramiñán Gilbert, et al. 2004. 'Legal and Ethical Framework for Astronauts in Space Sojourns: Proceedings'. European Center for Space Law. https://unesdoc.unesco.org/ark:/48223/pf0000139752.

Dyer-Witheford, Nick. 2004. '1844/2004/2044: The Return of Species-Being'. *Historical Materialism* 12 (4): 3–25. https://doi.org/10.1163/1569206043505130.

Elias, Norbert. 1978. *The History of Manners*. New York: Pantheon Books.

Graeber, David. 2007. 'Manners, Deference, and Private Property: Or, Elements for a General Theory of Hierarchy'. In idem, *Possibilities: Essays on Hierarchy, Rebellion, and Desire*, 13–55. Oakland, CA: AK Press.

———. 2013. 'It Is Value That Brings Universes into Being'. *HAU: Journal of Ethnographic Theory* 3 (2): 219–43. https://doi.org/10.14318/hau3.2.012.

Hazell, Robert, and Bob Morris. 2017. 'If the Queen Has No Reserve Powers Left, What Is the Modern Monarchy For?' *Review of Constitutional Studies* 22 (1): 5–32.

Hochschild, Arlie Russell. 2012. *The Managed Heart: Commercialization of Human Feeling*. Berkeley: University of California Press.

James, Cyril Lionel Robert, Raya Dunayevskaya, and Grace Lee Boggs. 2013. *State Capitalism and World Revolution*. Oakland, CA: PM Press.

Jeevendrampillai, David. 2017. 'Failure as Constructive Participation? Being Stupid in the Suburbs'. In *The Material Culture of Failure: When Things Do Wrong*, edited by Timothy Carroll, David Jeevendrampillai, Aaron Parkhurst, and Julie Shackelford, 113–32. London: Bloomsbury Academic.

Jeevendrampillai, David, Victor Buchli, Aaron Parkhurst, Adryon Kozel, Giles Bunch, Jenia Gorbanenko, and Makar Tereshin. 2023. 'An Ethnography of an Extraterrestrial Society: The International Space Station'. In *The Routledge Handbook of Social Studies of Outer Space*, edited by Juan Francisco Salazar and Alice Gorman. Abingdon: Routledge. www.routledgehandbooks.com/doi/10.4324/9781003280507-38.

Kantorowicz, Ernst Hartwig. 1997. *The King's Two Bodies: A Study in Mediaeval Political Theology*. Princeton, NJ: Princeton University Press.

Krittanawong, Chayakrit, Nitin Kumar Singh, Richard A. Scheuring, Emmanuel Urquieta, Eric M. Bershad, Timothy R. Macaulay, Scott Kaplin, et al. 2023. 'Human Health during Space Travel: State-of-the-Art Review'. *Cells* 12 (40): 1–32. https://doi.org/10.3390/cells12010040.

Marx, Karl. 1992. *Early Writings*. Translated by Rodney Livingstone and Gregor Benton. Penguin Classics. London: Penguin Books.

Melvin, Leland. 2018. *Chasing Space: An Astronaut's Story of Grit, Grace, and Second Chances*. New York: Amistad.

Olson, Valerie. 2010. 'The Ecobiopolitics of Space Biomedicine'. *Medical Anthropology* 29 (2): 170–93. https://doi.org/10.1080/01459741003715409.

———. 2018. *Into the Extreme: U.S. Environmental Systems and Politics beyond Earth*. Minneapolis: University of Minnesota Press.

Parkhurst, Aaron, and David Jeevendrampillai. 2020. 'Towards an Anthropology of Gravity: Emotion and Embodiment in Microgravity Environments'. *Emotion, Space and Society* 35 (May): 100680. https://doi.org/10.1016/j.emospa.2020.100680.

Patarin-Jossec, Julie. 2020. 'The Politics of Heroes' Body: Ethnographying the Training of Foreign Astronauts in Russia'. *Corpus Mundi* 1 (2): 14–36. https://doi.org/10.46539/cmj.v1i2.12.

———. 2021. *La fabrique de l'astronaute: Ethnographie terrestre de la station spatiale internationale*. Europes, terrains et sociétés. Paris: Editions Pétra.

———. 2022. 'Manufacturing National Heroes: Cosmonauts and Post-Soviet Identity in Putin's Russia'. In *Global Nationalism: Ideas, Movements and Dynamics in the Twenty-First Century*, edited by Nicholas Michelsen and Pablo de Orellana, 173–93. London: World Scientific Europe. https://doi.org/10.1142/9781800611542_0010.

Peacock, Vita. 2013. 'We, the Max Planck Society: A Study of Hierarchy in Germany'. PhD thesis, London: University College London.

———. 2015. 'The Negation of Hierarchy and Its Consequences'. *Anthropological Theory* 15 (1): 3–21. https://doi.org/10.1177/1463499614564887.

Postone, Moishe. 1996. *Time, Labor, and Social Domination: A Reinterpretation of Marx's Critical Theory*. New York: Cambridge University Press.

Sahlins, Marshall, and David Graeber. 2017. *On Kings*. Chicago, IL: Hau Books.

Snyder, Benjamin H. 2019. 'The Tyranny of Clock Time? Debating Fatigue in the US Truck Driving Industry'. *Time & Society* 28 (2): 697–720. https://doi.org/10.1177/0961463X17701955.

Stuster, Jack. 2010. *Behavioral Issues Associated with Long-Duration Space Expeditions: Review and Analysis of Astronaut Journals Experiment 01-E104 (Journals): Final Report*. Johnson Space Center, Houston, TX: National Aeronautics and Space Administration.

———. 2016. *Behavioral Issues Associated With Long Duration Space Expeditions: Review and Analysis of Astronaut Journals Experiment 01-E104 (Journals) Phase 2 Final Report*. Johnson Space Center, Houston, TX: National Aeronautics and Space Administration.

Taylor, Frederick W. 1967 [1911]. *The Principles of Scientific Management.* New York: Norton Library.
Thompson, Edward P. 1967. 'Time, Work-Discipline, and Industrial Capitalism'. *Past & Present* 38: 56–97. https://doi.org/10.1093/past/38.1.56.
Uri, John. 2020. 'The Real Story of the Skylab 4 "Strike" in Space'. 16 November 2020. www.nasa.gov/history/the-real-story-of-the-skylab-4-strike-in-space/.
Warnier, Jean-Pierre. 2007. *The Pot-King: The Body and Technologies of Power.* Leiden: Brill. https://doi.org/10.1163/ej.9789004152175.i-325.
Wartenberg, Thomas E. 1982. '"Species-Being" and "Human Nature" in Marx'. *Human Studies* 5 (2): 77–95. https://doi.org/10.1007/bf02127669.
Weibel, Deana L. 2019. 'Astronauts vs. Mortals: Space Workers, Jain Ascetics, and NASA's Transcendent Few'. *Space Review*, 8 April 2019. www.thespacereview.com/article/3690/1.
———. 2024. 'Astronauts as Chosen People: Religious Ways of Understanding the Astronaut Experience and Life After Space'. In *Religion and Outer Space*, edited by Eric Michael Mazur and Sarah McFarland Taylor, 216–32. Abingdon: Routledge. https://doi.org/10.4324/9781003088264-18.
Wierzbicka, Anna. 1986. 'Human Emotions: Universal or Culture-Specific?' *American Anthropologist* 88 (3): 584–94. https://doi.org/10.1525/aa.1986.88.3.02a00030.
Wikan, Unni. 1989. 'Managing the Heart to Brighten Face and Soul: Emotions in Balinese Morality and Health Care'. *American Ethnologist* 16 (2): 294–312. https://doi.org/10.1525/ae.1989.16.2.02a00070.

3 Anthropologists in outer space

Science fiction, infrastructure, comparison

Chakad Ojani

Anthropologists sometimes feature as protagonists in science fictional accounts on space exploration. Ursula K. Le Guin's novels are emblematic examples. In these stories, the anthropologist serves as heuristic for speculative world-building, creatively laying the ground for reflections on ethics and the human condition. Encounters with difference on other planets work as a mirror that allows for generative contrasts and juxtapositions, not unlike the comparative effects that anthropologists have long sought to bring about through ethnographic immersion on Earth. At the same time, in recent years, anthropologists have also become increasingly drawn to science fiction as a source of inspiration for thinking across worlds (for example, Anderson et al. 2018; Jensen and Kemiksiz 2019).

Inspired by the way science fiction and anthropology bleed into one another, this chapter makes use of Adrian Tchaikovsky's novella *Elder Race*, which features an anthropologist, as a stepping stone for discussing the possibilities of extraterrestrial ethnography. Moreover, by drawing on fieldwork on outer space infrastructures and imaginaries in subarctic Sweden, the chapter intimates that the anthropology of infrastructures, by virtue of its topological approach to socio-material relations, is particularly apposite for attending to processes of co-emergence between off-Earth milieus and more-than-human sociality on the planet. Whereas the science fictional anthropologist-protagonist becomes a heuristic for making explicit earthbound qualities by situating anthropos in space, the anthropology of infrastructures helps foreground cosmic relations by showing how extraterrestrial processes are sometimes brought down to Earth.

I develop my argument over four sections, first by eliciting the science fictional qualities of anthropology through a discussion on Ursula K. Le Guin's *The Left Hand of Darkness*. Given its use of equivocations as a prompt for conceptual displacement and innovation, anthropological analysis holds much in common with science fictional world-building; both treat limits and limitations as a resource for creativity. Next, I turn to

DOI: 10.4324/9781003437956-3

Tchaikovsky's *Elder Race* to highlight additional links between the two genres. In his attempt to understand the local culture on Sophos 4, the novella's anthropologist-protagonist and lead character, Nyr, undergoes a transformation that bears striking resemblance to the shift from pre- to post-*Writing Culture* (Clifford and Marcus 1986) sensibilities toward ethnographic encounters. By further showing how Nyr is forced to confront the limits of anthropology wholesale, I suggest that the story gestures toward an additional turn reminiscent of the discipline's regard for recursive analysis. *Elder Race* paves the way for a discussion on my own fieldwork around the Swedish rocket launch infrastructure and its impact area outside the subarctic city of Kiruna, where the Swedish Space Corporation is currently developing orbital launch capability. I discuss how space folds into and reshapes terrestrial milieus in a number of ways, both infrastructurally and through my interlocutors' comparisons between space and underground resource extraction. A topological approach to infrastructural relations serves to bring such processes into view ethnographically, a point I discuss in the conclusion in terms of Earth's extraterrestrial-likeness.

Science fictional anthropology

Set on the planet Gethen, Ursula K. Le Guin's 1969 novel *The Left Hand of Darkness* recounts how a human native from Terra, Genly Ai, tries to persuade the nations of Gethen to join a confederation of planets called Ekumen. While not strictly an anthropologist, Ai's role as an envoy sent from another planet puts him in situations that are remarkably anthropological. Throughout the story, Ai struggles to understand social life on Gethen, which, in contrast to his own society, has been shaped by the absence of fixed gender characteristics. By placing Ai on this planet, Le Guin creates a science fictional encounter between worlds that prompts us to question deep-seated assumptions around sex and gender on Earth, in particular through the relationship between Ai and Estraven, a Gethenian politician with whom Ai gradually builds trust and acceptance across difference.

In an article published in *Current Anthropology*, Leon E. Stover (1973) described *The Left Hand of Darkness* as "[t]he most sophisticated and technically plausible work of anthropological science fiction," a category that "enjoys the philosophical luxury of providing answers to the question 'What is man?' " (Stover 1973, 472). In Stover's assessment, what qualified Le Guin's novel to fit so neatly within this category was its depiction of "the relationship of culture and biology" and, more specifically, how "an alien life form, on another planet, is given a pattern of cultural behaviour suitable to its extraterrestrial biology" (Stover 1973, 472).

In the introduction to a later edition of this novel, Le Guin herself framed science fiction more generally in terms that resonate strongly with anthropology's deep-seated concern with the otherwise. Contrary to what one might assume, she proposed that the genre deals first and foremost with the present, not the future. As a "thought-experiment" (Le Guin 2018, xviii), science fiction often explores the imagined consequences of select socio-technical trends through their fictional intensification, thus accentuating aspects of the present that usually remain backgrounded. Philosopher and cultural critic Steven Shaviro (2015) has recently framed science fiction along analogous lines. Rather than predicting the future, science fiction "works to *extrapolate* elements of the present, to consider what these elements might lead to if allowed to reach their full potential" (Shaviro 2015, 2; original emphasis). Science fiction, then, deals not so much with "the actual future but the *futurity* that haunts the present" or, phrased differently, "the *virtual* dimensions of existence" (Shaviro 2015, 2; original emphasis). Accordingly, *The Left Hand of Darkness* narrates an encounter with a society without men and women, thereby encouraging us to reconsider existing gender relations and imagine what other potentialities are prevented from actualisation.

Martin Savransky's (2016) characterisation of speculation as an exercise in creativity *through* – not in spite of – constraints is telling, of how science fiction and ethnography hold much in common. For Savransky, speculative thinking in social analysis "takes up the stubbornness of facts as a constraint upon its own creative activity" (Savransky 2016, 201–2). It accepts the limitations of a topic and sets out to "enlarge and recast the categoreal ideas within the limits of that topic" (Whitehead, cited in Savransky 2016, 199). Similarly, in science fictional thought experiments, the intensification of a contemporary trend becomes the basis for world-building. Pushed against its own limits, the given phenomenon is assessed on the basis of its envisaged consequences on other aspects of sociality, no matter how improbable those ramifications may be.

This bears uncanny resemblance to how anthropologists have treated encounters with the limits of ethnographic enquiry as an opportunity to displace their own pre-analytical assumptions and categories (Ojani 2022). Consider Martin Holbraad and Morten Axel Pedersen's (2017) explanation of Eduardo Viveiros de Castro's method of controlled equivocation: If an interlocutor assumes or says that "what both he and the anthropologist see as a gift is something that contains a spirit," and if this contention clashes with the anthropologist's assumption that the gift is an inanimate object, "then the anthropologist must ask himself reflexively: what is it about the way I define gifts that makes this native assumption appear incongruous? How do I need to change my definition in order to remove this intensional incongruity?" (Holbraad and Pedersen 2017, 188). Doing so, in turn, amounts to an "ever-precarious, complex and inherently experimental exercise of mutual calibration between ethnographic materials and their anthropological conceptualization," through which an array of concepts and associations might require reconsideration – not least "the operative distinction between spirits and objects," but then also " 'property', 'labour', 'fame', 'honour', 'profit' or 'self-interest' " and potentially the meaning of "transaction" as well as the relation between "action" and "event" (Holbraad and Pedersen 2017, 189). The initial equivocation between anthropologist and interlocutor becomes a constraint that demands a modulation of "the conceptual repertoire on which the anthropologist relies, so as to be able to *arrive* at sensible descriptions of the ethnographic objects in question" (Holbraad and Pedersen 2017, 194; original emphasis), thereby rendering the analysis an exercise that comes close to science fictional world-building described above; both tap into the limitations of a select observation to reconfigure broader conceptual or socio-material milieus.

Accordingly, if Stover (1973) found in the *Left Hand of Darkness* an example of what he called "anthropological science fiction," we might likewise speak of the science fictional qualities of *anthropology*. Consider, for instance, their shared "commitment to difference" (Anderson et al. 2018) and anthropologists' burgeoning regard for speculative fiction (Jensen and Kemiksiz 2019), the play with reality and fiction in *Body Ritual Among the Nacirema* (Miner 1956) or *The Teachings of Don Juan* (Castaneda 1968), the speculative inclinations in early anthropological writings on off-Earth settlement and extraterrestrial encounters (Maruyama and Harkins 1975), or, as hinted above, the inventive nature of anthropological analysis and ethnographic practice more generally (Criado and Estalella 2023; Wagner 1981; see also Jeevendrampillai and Fortais, Chapter 9, this volume). Indeed, as Casper Bruun Jensen and Asli Kemiksiz (2019) have commented, "In SF, people journey to different worlds, and foreigners arrive unexpectedly in 'ours.' Analogously, the anthropological imagination thrives on perspective shifts often brought about by traveling far from home" (Jensen and Kemiksiz 2019, iv). This section has thus moved from anthropological science fiction to what can be denoted *science fictional anthropology*; that is, an understanding of anthropology in which speculation is integral.

Anthropos in space

In Adrian Tchaikovsky's novella, *Elder Race*, we are introduced to Lyn, Fourth Daughter of the Queen of Lannesite, and her ally Esha of the Coast-people on their way to Tower of Nyrgroth Elder. The natives of this world understand Nygroth to be the last of the ancients, an Elder Race and sorcerer of legendary powers. A demon is said to terrorise a vassal state where the poor folks live, but the Queen has dismissed all such alarms as grounded in mere fantasy. Lyn is determined to prove everyone wrong by conquering the demon. Never having been taken seriously by the rulers of Lannesite, Lyn is on a quest to forever change the way she is regarded by asking Nygroth Elder to fight magic with magic.

But things are not as they seem. In the second chapter, after the two travellers are let in through the portal to the sorcerer's domain, Tchaikovsky switches to the perspective of Nygroth, after which the chapters jump back and forth between the two perspectives. Throughout the story, Lyn, Esha, and the other natives remain oblivious to the fact that Nygroth Elder is actually Nyr Illim Tetvitch, an anthropologist "second class" of Earth's Explorer Corps. We learn that he is on Sophos 4, one of the planets seeded by humanity's generation ships millennia ago and then left to evolve independently. Together with his colleagues, Nyr had been part of a later expedition to the colony with the aim to study its evolution. Alas, things started going wrong on Earth and his colleagues had since long returned. Cut off from any contact with Earth for centuries, Nyr was now left on his own, protected by various non-contamination regulations and sleeping time away to stop himself from growing old.

Until he learns what the "post-colonial natives" actually want, Nyr suspects they might have come to kill him "with salvaged old Earth tech" (Tchaikovsky 2021, 30–31). Lonely, depressed, and despairing over his status as "a very bad anthropologist" (Tchaikovsky 2021, 43) whose articles no one will read about a society he fails to understand, Nyr decides to give his visitors the benefit of the doubt and accompanies them on a journey to find out what the story about the demon is really about, albeit without taking it at face value. It is a chance to learn something new about a culture that he has never quite grasped – in part, he thinks, because of an earlier mistake of his to intervene in local affairs.

From this point on, we are thrown into a tale of mutual misunderstanding or what Viveiros de Castro might call "uncontrolled" equivocation (see Viveiros de Castro 2004, 5). Meanwhile, Nyr, in his obsession with "maximum objectivity" (Tchaikovsky 2021, 84), weighs every decision to interfere in local life against the risk of cultural contamination, all the while complaining about the locals' failure to understand him. Tchaikovsky has taken this disposition to its extreme by equipping the anthropologist with a Dissociative Cognition System (DCS) that shields him from emotions and allows complete rationality. The sense of awkwardness and not-quite-fitting-in that anthropologists find so generative are effectively shut out: "The DCS keeps my embarrassment and awkwardness at bay, and I pass through them with a neutral demeanour, as benefits an academic" (Tchaikovsky 2021, 58).

And yet, Nyr is deeply concerned about the "social scars" he has already caused and might be occasioning by setting out on the journey, making him "not only the last but the worst anthropologist" (Tchaikovsky 2021, 64). With his characteristic sense of self-pity, he therefore thinks he is "lucky that nobody is coming back to read [his] reports," for in that case he would "be on the first ship home" (2021, 64). By the same token, whenever he manages not to stand out and successfully becomes "part of a cultural script that these people understand" (2021, 65), Nyr expresses a sense of professional pride, and the same goes when he makes new discoveries. For example, at one point he suspects that he has been brought on a ritual venture, and through which Lyn tries to prove herself by "acting out some legend" (2021, 79). He wonders if she thinks "there is a literal beast or not," just as he is "not sure if her 'demon' is real or just symbolic" (2021, 82). To enquire into the matter would nonetheless be "patently taboo" (2021, 79), and so Nyr has no choice but to play along. Once he gets back, he will have "the mother of all reports to file" (2021, 83).

A central equivocation in the story pertains to the locals' perception of Nyr as a sorcerer, when according to the anthropologist no such thing can possibly exist: "They think I'm a wizard. They think I'm a fucking wizard. That's what I am to them, some weird goblin man from another time with magic powers. And I literally do not have the language to tell them otherwise" (2021, 85). Yet conversely, Nyr's behaviour does not always align with Lyn's understanding of how a sorcerer should behave. When Nyr expresses surprise about aspects of local custom that he had previously overlooked, Lyn finds it rare that "a sorcerer of the ancient race was taught something new" (2021, 71). Nyr tries in vain to set the record straight once and for all by finding a way around the fact that "scientist" and cognate terms all translate into "magician," "wizard," "sorcerer," or "magus." He just happens to know how the world works, he explains, and he makes use of his technology to bend the world according to his will. But, Lyn asks, "Is that not what magic is?" (2021, 109). In an act of frustration, Nyr finally admits that his duty as an anthropologist is to let them all die: "I should just walk away and let this happen, and record the story of it for those to come, as though there is anyone who will ever come. I should not be here. I am not part of your stories" (2021, 110).

In what is perhaps the most enthralling chapter of the book, Tchaikovsky juxtaposes the anthropologist's narrative and Lyn's interpretation side by side in two columns and exemplifies the equivocations even more vividly. When Nyr mentions "Earth," Lyn hears "the otherworld" (2021, 111), neatly fitting Nyr's account within the framework of an origin myth. The anthropologist's narrative about cultural degeneration and, moreover, about the natives of this planet as an outcome of colonisers' miscalculations about the capacity for self-sustenance, translates into a story about a loss of the ancients' magic but continuity of life and prosperity notwithstanding. Whereas for the locals Nyr is an "ancient," for Nyr the natives are mere "savages" far below his own "superior culture" (2021, 147), as it were. The protagonists see each other in ways that their respective other cannot understand. Roy Wagner might have concluded that "Their misunderstanding of me was not the same as my misunderstanding of them" (Wagner 1981, 20). A brief moment of regret over what he has just disclosed is remedied by Nyr's bittersweet recognition of this mutual misapprehension. Even when he had left his "professional integrity in tatters" (Tchaikovsky 2021, 116) by succumbing to frustration and revealing what kind of place Sophos 4 actually is, the natives ultimately did not have the slightest clue what he was going on about: "Somehow I told them something else instead" (2021, 116).

However, as the story unfolds, it takes an interesting turn that could very well mark a difference between pre- and post-*Writing Culture* (Clifford and Marcus 1986), when anthropologists began to reflexively interrogate the power relations underpinning their practices and claims on objectivity. At one point, Nyr ponders how he is supposed to explain that in case a battle against the supposed demon ends with his death, there is a satellite in orbit that will make sure his body is destroyed so as to prevent the natives from getting hold of it, thereby eliminating the risk of cultural contamination. When reflecting on this, he starts to question the ethics of his own anthropological practice: "They set us here to make exhaustive anthropological notes on the fall of every sparrow. But not to catch a single one of them. To *know*, but very emphatically not to *care*" (Tchaikovsky 2021, 148; original emphasis). Having trusted blindly in the unquestionable powers of the Elder, Lyn concomitantly begins suspecting that it is actually the sorcerer who fails to understand *her*, rather than the other way around. And when Nyr, Lyn, and their fellow travellers finally encounter the demon, it turns out that it is indeed beyond the grasp of Nyr's understanding. He simply cannot figure out what it is, which implies that there do exist phenomena in this world that escape the conceptual toolbox of his earthbound anthropology.

Back in his tower and in a move that is reminiscent of the regard for recursivity in contemporary anthropology, the story ends with the undoing of the discipline as Nyr has come to know it:

> And I decide, with my most rational mind, that I am no longer an anthropologist. My failures of objectivity and detachment surely meant that anything I wrote would be hopelessly contaminated by my involvement with the culture I purport to study. Similarly, this place is no longer an outpost. To be an outpost requires some larger thing to be posted out of, and I can be honest with myself: there is no larger thing;

> not for any practical intents and purposes, and most likely not at all in any way. This is nothing but a tower, and I am nothing but a scientist of sufficiently advanced technology, which is to say a magician.
>
> (Tchaikovsky 2021, 197)

Nyr finally accepts that he cannot dissect himself from the world he has set out to study. Through a narrative technique that is familiar from the aforementioned *Left Hand of Darkness*, Tchaikovsky masterfully situates anthropos in space and speculatively brings into view the limitations of Nyr's terrestrial categories. One might speculate that, had there been a sequel to this novella narrating his remaining time on the planet, Nyr would have replaced his inherited anthropology with a means to control the equivocations between him and his interlocutors.

Space on Earth

Anthropologist-protagonists in *Elder Race* and other such stories become a heuristic for bringing into view and reflecting on earthbound assumptions. The argument I now wish to pursue is that the anthropology of infrastructures offers conceptual and methodological resources for studying processes that run in the opposite direction (Timko et al. 2022, 10). I suggest that the topological approach to socio-material relations developed in this field is especially apposite for foregrounding how the extraterrestrial is occasionally brought down to Earth (see Harvey 2012).

In 2022, I embarked on an ethnographic fieldwork on outer space infrastructures and imaginaries in Sweden. I had been drawn to this topic a couple of years earlier, after having come across news that the state-owned Space Corporation was about to turn its sounding rocket range, Esrange, into a spaceport with small satellite launch capability. Situated some 40 kilometres outside the subarctic city of Kiruna and with an impact area that overlaps with the legally recognised reindeer-herding territories of four Sámi villages, the current expansion of Esrange should be understood against the background of the broader commercialisation of space and concomitant growth in demands on satellite data. In this context, the Swedish space industry is hoping to attract not only space agencies and the international scientific community, as has been predominantly the case to date, but increasingly also commercial actors in need of test beds and launch services. I hoped to elicit the "planetary imaginaries" (Messeri 2016) underpinning the expansion, but my plan was first and foremost to treat it as an avenue for investigating the role of space infrastructures in modifying orbital-planetary relations both conceptually and materially.

As Christine Bichsel (2020) has noted, human engagement with the extraterrestrial is often (but certainly not always) a "highly material and technology-intensive activity" (Bichsel 2020, 3), rendering enquiry into space infrastructure "a key entry point for unravelling the relationalities of Earth and outer space" (Bichsel 2020, 2; see also Chinigò and Nieber, Chapter 4, this volume). A by now well-established idea in the anthropology of infrastructures is that infrastructures are both "things and also the relation between things" (Larkin 2013, 329). Drawing on such framings, Jensen and Morita (2017) have further proposed that we understand infrastructures as "open-ended experimental systems" that "produce novel configurations of the world" (Jensen and Morita 2017, 618, emphasis omitted). By creating new kinds of relations between such diverse entities as states, spirits, nature, forms of knowledge, and people, infrastructures become more than scripts that can be ethnographically mined for cultural representations about, say, modernity or progress. This is only part of the story. Rather than simply mirroring social relations, infrastructures also modify them, including "what can be perceived as 'social' or 'natural' " to begin with (Jensen and Morita 2017, 618). Taking my cue from Bichsel's observations and the anthropology of infrastructures more broadly, in my research I was curious how thickening infrastructural relations between Earth and space might be reshaping the domains drawn together.

One example of such reconfiguration is the recent emergence of space weather and orbital debris as matters of concern (Clormann and Klimburg-Witjes 2021; Taylor 2020). Processes that surpass the threshold of human perception have suddenly emerged as things around which space professionals need to gather. These phenomena must be visualised, counted, tracked, predicted, and managed so as to mitigate the risks they pose to critical ground-based infrastructures such as data centres and systems that support global trade, commerce, and weather forecasting. In the processes, orbital space is increasingly conceptualised in environmental terms, meaning that post-terrestrial infrastructures transform what humans encounter *as* the environment as well as the boundaries "where environmental relations begin and end" (Olson 2018, 224). Space, then, is present in earthly activities not only through the way it is "useful" in our daily lives, as space actors in Sweden were often quick to remind me, but likewise through the way material processes unfolding far beyond the planet are mediated by the vulnerabilities of satellites in orbit. Indeed, talk about these vulnerabilities and hazards was ubiquitous in the many space conferences, seminars, and events I attended over the course of my fieldwork. Are there other emergent ways in which space is made present in subarctic Sweden?, I wondered.

Shortly upon my arrival in Kiruna, however, I was struck by the difficulty of conducting fieldwork on human engagements with a domain so remote and in an area so vast. Many of my interlocutors, especially the reindeer herders, engaged in activities that were scattered over a geographical setting twice the size of Luxembourg. In addition, access to key actors in the Swedish space sector turned out to be no easy feat. If the anthropologist in *Elder Race* studied "down," I was certainly studying "up" or "sideways" when engaging with some of these actors (Hannerz 2006), and my fieldwork was continuously obstructed by countless unanswered emails and phone calls and last-minute cancellations of meetings and interviews.

In the meantime, a serendipitous encounter with an employee at the local municipality – who, it turned out, had recently completed his undergraduate degree at the anthropology department where I myself had once studied – led to joint weekend trips to various active and long since inactive mines. These excursions, I suspect, was the result of two outsiders' attempt to act according to what we understood to be part of a local script; as the week drew to a close, many locals packed their gear and set out on their leisurely activities of choice outside the city. Despite his relatively short time in Kiruna, my newfound companion, Mattias, had acquired a relatively broad network of friends and acquaintances, partly owing to the nature of his job but also his friendly and outreaching personality. Mattias put me in touch with several local residents, one of whom even turned out to have held a job in the space sector before being laid off, after which he had done various odd jobs in the local mining industry – a connection between industries that would later prove central to my research.

Driving to various mining areas with Mattias in the 4X4 I was using for my fieldwork, and back and forth between my interlocutors' homes and workplaces, it gradually dawned on me that socio-technical engagements with space and the skyward in Kiruna had been preceded by a set of more long-running vertical relations extending downwards. In fact, a main motivation for placing the rocket range in this area to begin with was the accessibility afforded by an infrastructure that had evolved around mining at least since the late nineteenth-century establishment of the Kiruna mine, one of the world's largest underground iron ore mines. And if the latter had consolidated a vertical territorial understanding with respect to the subterranean, the contemporary expansion of the launch site relies on a similar understanding albeit with respect to the atmospheric. As it happens, one of the Swedish Space Corporation's main selling points is the region's relatively unoccupied airspace.

Moreover, in my conversations with local residents, I soon realised that space in Kiruna was often made sense of by way of comparison with mining and the underground. Several of my interlocutors pointed out to me that Kiruna is not really a space town, as space actors would often have it, but rather a mining town. At times, they would ascribe equal weight to both sectors by maintaining the primacy of mining for Kiruna's identity to date, yet underscoring the space industry's importance for the city's more distant and unpredictable future. Space system engineers, space lobbyists, and others working in the Swedish space sector invoked such relations too. For instance, in arguing for possible synergies between the local mining and space industries, they often drew connections between the two domains by declaring that both are "high tech" and deal with "extreme environments," hence offering an analogy between distinct environmental domains familiar from other ethnographic settings (for example, Helmreich 2009; Olson 2018).

One such comparison was particularly captivating, made by a PhD student in space systems at the local Space Campus. As was common in my encounters with space actors, the conversation had gradually segued into the topic of space colonisation, which my interlocutor considered inevitable but as a by-product of off-Earth mining. Perhaps noticing my surprise, he illustrated his point by orienting my attention to how Kiruna turned into the city it is today only as a consequence of the local iron ore mine. A similar process would unfold in space, he insinuated, meaning that the city's relation to mining served to envisage human settlement in space. By the same token, space gained meaning by analogy to Kiruna's history of underground resource extraction. My observation above that space infrastructures fold the extraterrestrial into ground was an etic interpretation regardless of "Indigenous" understandings. By contrast, now I came to realise that space was also brought down to Earth *emically* through analogy's capacity to enlarge the meanings of its respective terms by transposing them into the context of the other (Strathern 2006).

Our weekend trips to mining pits in the city's surrounds suddenly gained new meaning, no longer disconnected from my study on space infrastructures. If the launch infrastructure required that the landscape be constructed as an "empty" impact area, then this had been made possible by a longer history of underground resource extraction. While frequently drawn upon to speculate on the futures of, and in, space, mining also predicated the envisaged emergence of those future scenarios.

What can space do?

In a chapter on extraterrestrial methods, Victor Buchli suggests that "the study of the extraterrestrial reconfigures our understanding of terrestrial realms in a profound way;" for example, by revealing "how extraterrestrial worlds are made Earth-like, and how Earth is made extraterrestrial-like" (Buchli 2020, 22). This chapter has deployed a somewhat different vocabulary. Whereas *Elder Race* situated the earthbound in space, in the context of my fieldwork, the extraterrestrial was brought down to Earth. Borrowing from Peter Sloterdijk (2009), we might think of these as acts or moments of "explication," whereby the introduction of alien elements in a new milieu renders perceptible things formerly inconspicuous. In the process, space becomes Earth-like and Earth extraterrestrial-like (Szolucha et al. 2023, 10–11). But what does this actually mean?

In *A Foray into the Worlds of Animals and Humans*, biologist Jakob von Uexküll (2010) used a similar idiom when describing the relation between spiders, their webs, and flies. As he explained, the spider's web contains fly-like qualities in its very design because it has been crafted to entrap what the spider is unable to anticipate with its own, unassisted senses. The fly's lifeworld is actively transcoded into the web by the spider, which further means that, as an extension of its body, the web becomes a conduit for fly-like features to leak into the spider's bodily constitution. Timothy Morton's (2013) account of how gamma rays "tune to us" is illustrative of how such processes might also unfold in other contexts. Morton notes that "It is very hard to see a gamma ray in itself. You have to cause it to be deflected in some way, or to mark some inscribable surface such as a photographic plate" (Morton 2013, 33). Perceiving gamma rays involves the introduction of a foreign element that brings the gamma rays into visibility vicariously through their traces. Accordingly, at the quantum level, "to see something just is to hit it with a photon or an electron: hence to alter it in some way. Every seeing, every measurement, is also an adjustment, a parody, a translation, and interpretation. A tune and a tuning." Phrased differently, "Gamma rays tune to us, gamma ray-pomorphizing us into a gamma ray-centric parody of ourselves" (Morton 2013, 33).

This amounts to a radical reconsideration of what gamma rays can do. Akin to how the fly "fly-pomorphises" the spider into a fly-centric parody, turning the spider fly-like, in Morton's example scientists are made gamma ray-like. It follows that we can learn something about flies by looking at spiders and their webs, and potentially about gamma rays by studying how experimenters, their tools, laboratories, and conceptual repertoires become gamma ray-like. It is against this background that my above account should be understood and, by extension, anthropologists' uses of speculative fiction for thinking across partially connected worlds. The space-likeness of Earth and Earth-likeness of space is suggestive of potential "contact zones" (Haraway 2008) between things seemingly out of touch with one another.

With respect to space infrastructures, the ethnographic question is: how are thickening infrastructural relations to the extraterrestrial reconfiguring or un-earthing the seemingly earthbound? In other words, how is space made to appear on Earth, and what can it do?

Acknowledgements

This research was supported by Riksbankens Jubileumsfond and the National Science Centre in Poland, project number 2020/38/E/HS3/00241.

References

Anderson, Ryan, Emma Louise Backe, Taylor Nelms, Elizabeth Reddy, and Jeremy Trombley. 2018. "Introduction: Speculative Anthropologies." *Theorizing the Contemporary, Fieldsights* December 18. https://culanth.org/fieldsights/introduction-speculative-anthropologies.

Bichsel, Christine. 2020. "Introduction: Infrastructure on/off Earth." *Roadsides* 3: 1–6. https://doi.org/10.26034/roadsides-202000301.

Buchli, Victor. 2020. "Extraterrestrial Methods: Towards an Ethnography of the ISS." In *Lineages and Advancements in Material Culture Studies: Perspectives from UCL Anthropology*, edited by Timothy Carroll, Antonia Walford, and Shireen Walton, 1st ed., 17–32. Abingdon: Routledge.

Castaneda, Carlos. 1968. *The Teachings of Don Juan: A Yaqui Way of Knowledge.* Berkeley: University of California Press.

Clifford, James, and George E. Marcus (eds). 1986. *Writing Culture: The Poetics and Politics of Ethnography.* Berkeley: University of California Press.

Clormann, Michael, and Nina Klimburg-Witjes. 2021. "Troubled Orbits and Earthly Concerns: Space Debris as a Boundary Infrastructure." *Science, Technology, & Human Values* 47 (5): 960–85. https://doi.org/10.1177/01622439211023554.

Criado, Tomás Sánchez, and Adolfo Estalella (eds). 2023. *An Ethnographic Inventory: Field Devices for Anthropological Inquiry.* Abingdon: Routledge.

Hannerz, Ulf. 2006. "Studying down, up, Sideways, Through, Backwards, Forwards, Away and at Home: Reflections on the Field Worries of an Expansive Discipline." In *Locating the Field: Space, Place and Context in Anthropology*, edited by Simon Coleman and Peter Collins, 23–41. Oxford: Berg.

Haraway, Donna J. 2008. *When Species Meet.* Minneapolis: University of Minnesota Press.

Harvey, Penny. 2012. "The Topological Quality of Infrastructural Relation: An Ethnographic Approach." *Theory, Culture & Society* 29 (4–5): 76–92. https://doi.org/10.1177/0263276412448827.

Helmreich, Stefan. 2009. *Alien Ocean: Anthropological Voyages in Microbial Seas.* Berkeley: University of California Press.

Holbraad, Martin, and Morten Axel Pedersen. 2017. *The Ontological Turn: An Anthropological Exposition.* Cambridge: Cambridge University Press.

Jensen, Casper Bruun, and Asli Kemiksiz. 2019. "Experiments in Thinking Across Worlds: Introduction." *NatureCulture* 5: i–xiii. www.natcult.net/journal/issue-5/experiments-in-thinking-across-worlds/.

Jensen, Casper Bruun, and Atsuro Morita. 2017. "Introduction: Infrastructures as Ontological Experiments." *Ethnos: Journal of Anthropology* 82 (4): 615–26. https://doi.org/10.1080/00141844.2015.1107607.

Larkin, Brian. 2013. "The Politics and Poetics of Infrastructure." *Annual Review of Anthropology* 42: 327–43. https://doi.org/10.1146/annurev-anthro-092412-155522.

Le Guin, Ursula K. 2018. *The Left Hand of Darkness*. London: Gollancz.

Maruyama, Magoroh, and Arthur Harkins (eds). 1975. *Cultures Beyond Earth: The Role of Anthropology in Outer Space*. New York: Random House.

Messeri, Lisa. 2016. *Placing Outer Space: An Earthly Ethnography of Other Worlds*. Durham, NC: Duke University Press.

Miner, Horace. 1956. "Body Ritual Among the Nacirema." *American Anthropologist* 58 (3): 503–7. https://doi.org/10.1525/aa.1956.58.3.02a00080.

Morton, Timothy. 2013. *Realist Magic: Objects, Ontology, Causality*. Ann Arbor, MI: Open Humanities Press.

Ojani, Chakad. 2022. "Speculative Relations in Lima: Encounters with the Limits of Fog Capture and Ethnography." *HAU: Journal of Ethnographic Theory* 12 (2): 468–81. https://doi.org/10.1086/720367.

Olson, Valerie. 2018. *Into the Extreme: US Environmental Systems and Politics Beyond Earth*. Minneapolis: University of Minnesota Press.

Savransky, Martin. 2016. *The Adventure of Relevance: An Ethics of Social Inquiry*. London: Palgrave Macmillan.

Shaviro, Steven. 2015. *No Speed Limit: Three Essays on Accelerationism*. Minneapolis: University of Minnesota Press.

Sloterdijk, Peter. 2009. *Terror from the Air*. Los Angeles, CA: Semiotext(e).

Stover, Leon E. 1973. "Anthropology and Science Fiction." *Current Anthropology* 14 (4): 471–74. https://doi.org/10.1086/201359.

Strathern, Marilyn. 2006. "Useful Knowledge." *Proceedings of the British Academy* 139: 73–109. https://doi.org/10.5871/bacad/9780197263945.003.0004.

Szolucha, Anna, Peter Timko, Chakad Ojani, and Karlijn Korpershoek. 2023. "Ethnographic Research of Outer Space: Challenges and Opportunities." *Ethnography* 0 (0): 1–20. https://doi.org/10.1177/14661381231220273.

Taylor, Alexander R. E. 2020. "Space Weather as a Threat to Critical Infrastructure." *Roadsides* 3: 63–72. https://doi.org/10.26034/roadsides-202000309.

Tchaikovsky, Adrian. 2021. *Elder Race*. New York: Tordotcom.

Timko, Peter, Karlijn Korpershoek, Chakad Ojani, and Anna Szolucha. 2022. "Exploring the Extraplanetary: Social Studies of Outer Space." *Anthropology Today* 38 (3): 9–12. https://doi.org/10.1111/1467-8322.12726.

Viveiros de Castro, Eduardo. 2004. "Perspectival Anthropology and the Method of Controlled Equivocation." *Tipití: Journal of the Society for the Anthropology of Lowland South America* 2 (1): 3–22.

Von Uexküll, Jakob. 2010. *A Foray into the Worlds of Animals and Humans: With a Theory of Meaning*. Minneapolis: University of Minnesota Press.

Wagner, Roy. 1981. *The Invention of Culture*. Chicago, IL: University of Chicago Press.

4 Imaginaries of outer space from Africa

Astronomy infrastructure in South Africa and Madagascar

Davide Chinigò and Hanna Nieber

The Square Kilometre Array (SKA) is an internationally funded effort to build the world's largest radio telescope infrastructure and advance research in radio astronomy, a branch of astronomy interested in studying celestial objects by detecting and analysing their radio frequency emissions. While dispersed across the globe through a myriad of research collaborations in astrophysics,[1] the majority of the telescope infrastructure is currently being built in South Africa. In this chapter, we draw on the ethnographic research conducted in South Africa and Madagascar to explore the imaginaries that are elicited in outer space research by looking at how SKA continental initiatives reverberate locally. We trace the roots of historical aspirations for outer space research and juxtapose them with aspirations for modernity based on the political discourse of the African Renaissance. We aim to query how aspirations materialise in real infrastructural projects, shaping physical landscapes and becoming entangled in local politics, as well as what happens when their materialisation remains potential.

Over the past decade, Africa has made its move into space, and the SKA is part of this broader trend. Public and private investments in space science and the space industry have surged dramatically (Pović et al. 2018). Earth-based space infrastructure, such as telescopes, have garnered particular traction. With its dark skies, areas of relative radio quietness, and a privileged view over the less explored Southern Hemisphere sky, the African continent offers sought-after locations for the placement of telescopes from the perspective of scientific research. Mediating our knowledge of outer space, the ever improving telescopes continue to excite the astrophysical community and elicit hopes for new data that will help understand ever more of the universe's dynamics. More broadly, telescopes prompt "the enthusiasm of imagination" (Mrázek 2002, 166), configuring outer space as a trope that symbolises modernity, internationalism, and a technologically driven "better" future.

DOI: 10.4324/9781003437956-4

However, as infrastructures, telescopes are not erected in a vacuum. In the words of Susan Leigh Star and Karen Ruhleder, grasping infrastructure requires dealing with their relationality, since they never emerge as things "stripped of use" (1996, 113). Telescopes are material structures that emerge across time and space in relation to institutions, legal policies, and land use regimes (Anand, Gupta, and Appel 2018; Edwards 2003; Mitchell 2002). The infrastructures themselves, and their anticipation, give shape and are shaped by everyday human experiences of hope, inclusion, and abandonment – they filter people's ties to a place through their visions for immediate futures. Entrenched in historical power relations and terrestrial conflicts around resource management, telescopes bring outer space imaginaries in touch with geopolitically informed conditions.

But how to study such a big infrastructural project ethnographically? How to acknowledge the aspiration that infrastructures elicit without resorting to a language of modernity that lends itself to justify land appropriation and inequality? How to engage with a topic that brings multiple scales, spaces (even outer space), and temporal frameworks into relation with each other? As this chapter will show, astrophysics in Madagascar cannot be detached from South Africa and historically situated notions of "Africa" that engulf Madagascar into South Africa's visions for radio astronomical infrastructure and the African continent as a whole. With this chapter, we point to the potentials of collaborative scholarly work between the fields of anthropology and history across time and space. Collectively, we demonstrate how astrophysical infrastructures on earth are concurrently shaped by situated local practices and imaginations, and by larger-scale geopolitical conditions and imaginaries of outer space.

To do this, we draw on our two ethnographic encounters with the SKA radio telescope in South Africa and Madagascar to explore the stratified layers of history and politics that inform aspirations and hopes that come with Africa's move to space. The SKA infrastructure is what ties our two fieldwork projects together. While in South Africa the precursor infrastructure (the MeerKAT telescope) to the international project is a pre-existing structure, in Madagascar the SKA is still at a planning stage. As noted by Brian Larkin, "what distinguishes infrastructures from technologies is that they are objects that create the grounds on which other objects operate, and when they do so they operate as systems" (Larkin 2013, 329). We show how the SKA as a pan-continental system also becomes the grounds for anticipatory prospects of astrophysical infrastructures in Madagascar, which are tied to ideological discourses and funding in South Africa. Engaging with studies in the anthropology of infrastructure (Star 1999; Larkin 2013; Anand, Gupta, and Appel 2018), we argue that approaching outer space through the lens of these situated terrestrial infrastructure projects allows us to show how national aspirations and commercial interests travel and transform between infrastructurally linked places. We do so in three steps.

First, we describe how space infrastructures gain traction through a political narrative centred on the role of progress in science and technology to fulfil the expectations of modernity raised by the post-colonial and post-apartheid transition in South Africa (Beinart and Dubow 2021). We begin our discussion by tracing the origin of a stream of political thought, the African Renaissance, which was key for gearing up the political momentum and investments in astronomy in South Africa in the aftermath of apartheid and for turning South Africa into a global exemplar in radio astronomy.

Second, we discuss how the scientific imaginary of outer space attached to the African Renaissance legitimated the repurposing of a vast area in the semi-arid Northern Cape of South Africa to build the telescope infrastructure, and how this clashed with other uses of this terrestrial space. We address this by attending to the expectations of modernity that accompany the development of outer space science in these locales, as well as the inherent tensions that emerge from the actual grounding of such imaginaries (Ferguson 1999; Archambault 2012; see also Anand 2012; Barry 2001; Mitchell 2002, 2011; Von Schnitzler 2008).

Third, we turn to Madagascar. Astrophysics' infrastructures bear aspirational values, desire, fantasy, pride. Larkin calls this "poetics" and notes that "infrastructures also exist as forms separate from their purely technical functioning [...] They emerge out of and store within them forms of desire and fantasy and can take on fetish-like aspects that sometimes can be wholly autonomous from their technical function" (Larkin 2013, 329). The international SKA project is envisioned to unfold across two consecutive phases, the first of which is currently implemented in South Africa. The second phase, the planning and implementation of which remains uncertain to this day, forecasts the expansion of the South African infrastructure to other sites across eight countries in southern, eastern, and western Africa, including Madagascar. Taking the perspective of young astronomy-enthusiasts in Madagascar, this final section examines the already emerging poetics of outer space infrastructure prior to its materialisation. It shows how desire, aspiration, and pride of outer space research shape imaginaries of terrestrial spaces, international collaborations, and personal senses of not being disconnected. It furthermore traces how, with the prospect of astronomy infrastructures emanating from South Africa, "Africa" becomes an epistemic object of hope. Such poetics of outer space infrastructure in Madagascar nurture personal relations to objects that may or may not materialise in the future and, in doing so, allow people to articulate their aspiration for a better future.[2]

The hypervisibility of space infrastructure and the African Renaissance

The SKA is led by the SKA Observatory, an inter-governmental organisation established by an international treaty signed in 2019.[3] The project encompasses a network of research collaborations across the globe – backed by national governments' commitments to a long-term budget.

While ideas to design the SKA date back to the 1990s, it was during the 2000s and early 2010s that institutional and research collaborations geared up towards implementation. In 2008, a project office, which later became the project's global headquarters, was established in the UK, and in 2011 the SKA was registered as a UK non-profit organisation. After a competitive bid, the decision was made in 2012 to split the planned physical infrastructure and create two telescopes, one in South Africa and the other in Australia. The decision ended a protracted competition for securing the international project between these two countries, one which had received extensive local and international media coverage. South Africa's competitiveness came as a surprise for many international observers, considering the much longer history of radio astronomy in Australia.

In South Africa, the decision was celebrated not only as an accomplishment for the domestic scientific community but also as a success for the country and the entire African continent. South Africa's inclusion of eight "African partner countries" helped to make the bid more convincing by framing the telescope as *African*, delineating "Africa" as a particular epistemic object that has political currency. As Naledi Pandor, the South African Minister of Science and Technology, commented:

> What am I feeling? I am excited, I am happy for our scientists, I am happy for our country, I am happy for Africa. We have done it! Who would have thought, so I am thrilled. [...] As a country and as the African partners we remain committed to the SKA project. We have always said we are ready to host the SKA and the world has listened to us.[4]

The SKA was deemed to be a tangible expression of "Africa rising," the continent's increasing participation in the global knowledge economy, and the fourth industrial revolution. Commenting on the SKA, former President of South Africa Jakob Zuma stated: "South Africa is confident that the country will deliver on the expectations of the continent and world."[5] In the post-apartheid moment, outer space research epitomised the expectations of modernity (Ferguson 1999; Redfield 2000), formulated along with science's claim to universality (see Nieber 2024). In conjunction with the SKA bid, a wider interest in radio astronomy emerged in South Africa, which today is coordinated under the South African Radio

Astronomy Observatory (SARAO) facility of the National Research Foundation.

The pursuit of optical astronomy in South Africa has a much longer history, dating back to the colonial period with the founding of the Royal Observatory in 1820 by the British. As noted by historian Saul Dubow (2019) in his historical review of astronomy's development in South Africa, during this period astronomy played an important practical and symbolic role in fulfilling the colonial civilisational mission (see also Osbourne, Chapter 5, this volume). From a practical perspective, astronomy contributed to advancements in navigation, cartography, and timekeeping, which were important for exploration and colonial endeavours. Astronomy, along with other sciences, was thus utilised by European powers to bolster their claims of superiority and justify the colonial project. The transition to apartheid marked an attempt by the white minority government to institutionalise astronomy as a national endeavour, especially during the 1980s, a period in which South Africa was isolated internationally (Dubow 2019).

With the end of apartheid in 1994, the African National Congress (ANC) government embraced and largely rebranded astronomy from a colonial and apartheid endeavour into a new force of transformation as part of the new political elite's continental and global ambitions. Under the presidency of Thabo Mbeki from 1999 to 2008, astrophysics and other key natural sciences assumed a new role in relation to discussions around the "African Renaissance." This line of political thought, inspired by pan-Africanism and humanism in the heydays of decolonisation (see Grilli and Gerits 2020), saw the progressive force of science as the catalyst by which African countries could finally break away from the legacy of colonialism (see Mavimbela 1998).

Against the backdrop of these ideas, it is at the beginning of the 2000s that efforts were made to build South Africa into a global exemplar in radio astronomy. South Africa initially joined the SKA with the status of observer in mid-2001. In 2005, the National Research Foundation (NRF) submitted a proposal to host the SKA infrastructure in the semi-arid and sparsely populated Northern Cape province together with selected sites in "African partner countries." The South African government committed significant financial and human capital investments to design and build two SKA precursors – KAT-7 and then MeerKAT. In 2009, the Department of Science and Technology inaugurated the site in the Northern Cape that was identified as the most suitable for radio astronomy (more below). In 2010, their open call for international projects made MeerKAT known among the global scientific community. South Africa's winning bid in 2012 to host a majority of the international infrastructural project marked the culmination of years of political discourse in which developments in astronomy and its infrastructures became the material proof of South Africa's advances, "bringing about a science Renaissance across the continent."[6]

Contrary to Star's (1999) contention that infrastructures are taken for granted and invisible "by definition," only to become visible when they break down, the SKA is an example of a hypervisible infrastructure rich with symbolism of the past and future of South Africa. The SKA infrastructure marks a shift in the narrative of astronomy as a colonial science – one that legitimates the colonial civilisational mission – to a science of transformation in the post-apartheid period. Similar to the other large infrastructural projects of that period, such as the Volta Dam in Ghana (Miescher and Tsikata 2009) or the Grand Renaissance dam in Ethiopia (Belay et al. 2020), the SKA carried symbolic value for nation-building aspirations through its hypervisibility that made the infrastructure possible and meaningful (cf. Barker 2005).

When in 2016 the first images produced by MeerKAT were released to the press after yielding scientific results far exceeding any initial expectations (Patel 2016), Pandor declared:

> The MeerKAT telescope, which is predominantly a locally designed and built instrument, shows the world that South Africa can compete in international research, engineering, technology and science. We are proud of our scientists and engineers for pioneering a radio telescope that will lead to ground-breaking research.[7]

The SKA's role was not only to be an infrastructure for astrophysical science, but also to extend and promote important social and economic benefits of science and technological development. Through scientific outer space imaginaries, terrestrial spaces and their socio-economic conditions were to be re-configured and the "African Renaissance" – including its conceptual work on "Africa" as a discursive construct – could flourish.

Expectations of modernity: contested terrestrial spaces

The core of the SKA telescope infrastructure is located in the semi-arid and sparsely populated central Karoo region of South Africa, 700 km north-east of Cape Town (see also Walker, Chinigò, and Dubow 2019). Since the end of the colonial period, the region's economy has centred around large-scale commercial sheep farming. Carnarvon, one of the nearest towns to the core site of the telescope, some 80 km away, has been designated as the "home of the SKA project." Counting approximately 5,000 residents, Carnarvon has high rates of unemployment and school dropout. It is an Afrikaans-speaking area, where over 90 per cent identify as "coloured" and 10 per cent as "white." Despite being a minority, the latter dominate Carnarvon's commercial sheep farming industry and the few other businesses based in town – an enduring legacy of apartheid (see also Chinigò 2019).

The physical presence of the SKA in the area around Carnarvon dates back to 2008, when the National Research Foundation acquired two farms to serve as construction sites for the KAT-7 radio telescope. Radio astronomy requires, on the one hand, low radio frequency interference and, on the other, good access to transport and logistical infrastructure, and this sparsely populated area was identified as well suited to these needs. Initially, and for about a year, the two farms retained grazing rights in the vicinity of the SKA and were allowed to continue their farming activities whilst the telescope's construction was still underway. This was an important moment, because it signalled the possibility for coexistence of farming and astronomy – flocks of sheep would hide from the sun in the shade under the KAT-7 dish (Figure 4.1). For the farmers living around the SKA, such possibility for coexistence symbolised the seamless

Figure 4.1 Sheep under the dishes.

Source: Carnarvon resident.

materialisation of the transition to a type of modernity that did not require any particular breakaway with the past. Rather, it was a type of modernity that sprouts out of local conditions and existing land uses. Sheep seemed to be thriving under the shade of the telescope. Residents perceived the prospect of the SKA as a valuable opportunity to rejuvenate a declining agricultural economy in one of the most marginalised regions of South Africa. In many ways, at this stage the SKA symbolised the materialisation of the expectations of modernity opened by the African Renaissance's call for a progressive role of modern science to pave the way to a new prosperous era.

As the KAT-7 project evolved into MeerKAT, however, the radio astronomy requirements to limit radio frequency interference around the infrastructure increased. In the following years, no farming was allowed on the premises of the telescope dishes. As the need for radio silence around the telescope increased, farming activities – for instance, the use of petrol-powered cars, cell phones, and other radio-emitting devices employed in modern farming – proved to be incompatible with the project's pursuit. The separation between sheep and dishes happened at the same time as fractures between Carnarvon residents and SKA management started to emerge. The narrative of the SKA as the driver of the African Renaissance changed and, with increasing restrictions in the area around the SKA site, critical voices gained volume. The compulsory Land Acquisition Programme (LAP) put in force between 2016 and 2017 to constitute a buffer zone against radio interferences around the core site was the initiative that attracted the most controversy. The LAP builds on a set of legislative initiatives that, starting with the 2007 Astronomy Geographic Advantage Act, have led to the declaration of the Karoo Central Astronomy Advantage Areas. These consist of three concentric areas of regulation spanning most of the Northern Cape Province and grow increasingly restrictive towards the core (see Figure 4.2). These legislations are meant to limit radio frequency interferences around the telescope infrastructure, and hence aim to prioritise astronomy over other human activities.

As well as the two mentioned above, 32 additional farms were subsequently purchased. While legally farmers were left with no option but to sell the land, anecdotal evidence highlights that compensations were quite generous and above the average market price. Two main grievances were voiced by the white commercial farmers' community. First, farmers complained that the land acquisition was imposed from above, and that when the SKA first approached Carnarvon, there was no mention about land acquisition beyond the two farms purchased by the project initially. Second, farmers argued that the economic impact of subtracting a huge tract of land from their local economy, primarily based on agriculture,

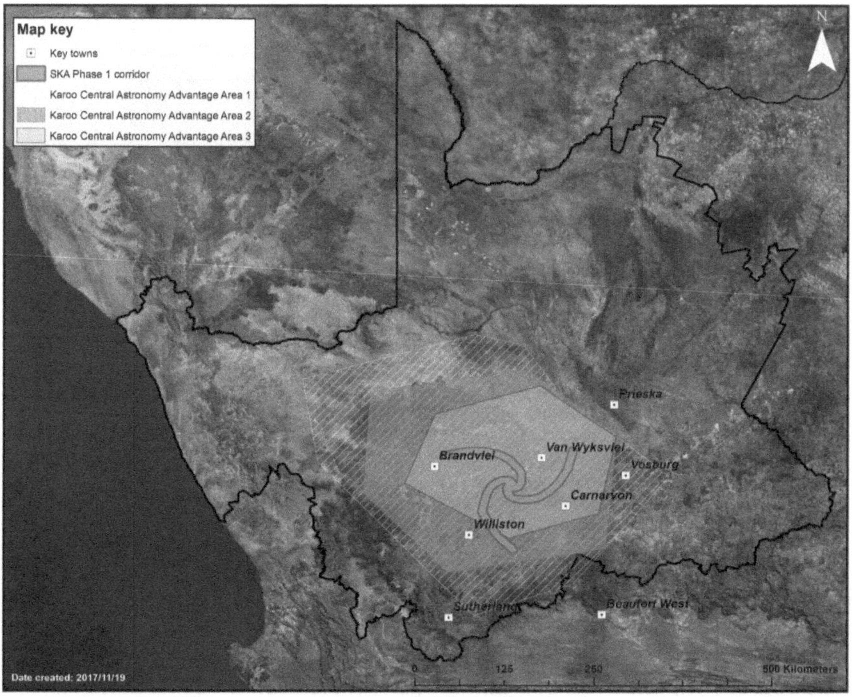

Figure 4.2 Map prepared by South African Research Chair in the Sociology of Land, Environment and Sustainable Development, Stellenbosch University.

was being underestimated by the legislators. A broader feeling of mistrust for local authorities underlined the white farmers' criticism of the land reform, given the role of land expropriation as a political tool used by the ANC national government to address the legacy of apartheid.

White farmers usually voiced these concerns in local media and in local engagement meetings organised by project management. As explained by one farmer, their main task was to challenge "the dominant narrative presented by the SKA and the government to the outside world that the SKA is constructed in a desert without people."[8] The presumption of an "empty" hinterland erased the farmers and needed to be vocally contested. Another white farmer confirmed that despite the initial promises, "it is now clear that the SKA is not acting in the interest of the local communities and what makes farmers angry is that the Karoo is presented as an empty land. Our estimations show that the project will have an impact on 46,000 people."[9] While probably the most cohesive and organised group, white farmers were only one among many voices in the local community questioning the SKA's adverse economic and environmental impact.[10]

In February 2017, SKA SA (now SARAO) signed an agreement with the farmers' organisation. The agreement was meant to explore "ways [in which] affected agricultural land is optimised to accommodate ongoing farming activities where possible, as long as the functioning of the radio observatory is not compromised" (SARAO 2017). Subsequently, the management of the 132,000ha of land that constituted the core of the SKA site in the Karoo was handed over to the South African National Parks (SANparks). Under its management, this core area became a "special nature reserve" and was renamed Meerkat National Park (ROSA 2020). Effectively, this is now a conservation area characterised by a special status that restricts access for all persons not involved in scientific research. No longer a farming area, occasional guided tours to the telescope site organised by SARAO local staff are the only possibility for local residents to visit the area.

The African Renaissance pursuit of astronomy in the name of scientific progress was marked by profound contradictions and historically loaded discussions around land use and land reform. The initial imperial vision behind the undertaking of astronomy under British colonialism was re-adapted twice. During apartheid, it was first elaborated in terms of a nationalist agenda (see Dubow 2019). Then, with the advent of democracy in 1994, the pursuit of astronomy was reconfigured as an important part of the African Renaissance. In this context, white farmers, the group who benefited from apartheid inequalities, have currently been displaced in the name of the pursuit of modern science. This allows us to draw important insights to the anthropology of infrastructure. First, the modernist vision of the African Renaissance justified the construction of the SKA as a symbol of the new democratic South Africa. This is an example of an infrastructure that elicits enthusiasm for a technologically mediated better future, a future that is also "just" because it is aimed at redressing past inequalities. Hypervisibility is therefore a characteristic of infrastructures for the exploration of outer space such as the SKA. While hypervisibility emphasises the future possibilities that infrastructures unlock, it tends to obscure the complex and contradictory histories within which infrastructures are enmeshed. Second, the restriction of other land uses required by radio astronomy, and the displacement generated by the SKA telescope in the Karoo, produced a significant experience of abjection at both individual and societal levels. Such experience is particularly prevalent when hypervisible infrastructures seemingly fail to fulfil the expectations of modernity that made them possible initially (Archambault 2012; Ferguson 1999). In Carnarvon, the grand narrative of "Africa rising" shows its price and these tensions are growing with time.

In Madagascar, where the construction of radio astronomy infrastructures has not yet commenced, the hopes for scientific progress still trump all worries about possible adversary effects.

The poetics of astronomy infrastructure: Malagasy potentialities

> Our professor said that Madagascar takes part in the SKA project. We don't know how it is going or what has happened. And it's already three years or more since he told us. [...] And this [vague talk about the SKA in Madagascar] is also the reason why many young people choose astronomy, because they are very interested in the SKA project and we would love to have a bit of a follow-up. According to the information we received, the telescope in Arivonimamo will be transformed into a radio telescope. It's a good future for Madagascar if we manage to complete this major project, because it's a very big project.[11]

This is a quote from an interview with Ilo, an astronomy student from Madagascar. He finished his physics "licence" (bachelor's degree) with distinction and is now in his first year of the astrophysics Master's programme. Ilo speaks excitedly about Madagascar's involvement in the Square Kilometre Array (SKA) project, for which he understands the conversion of a dish in Arivonimamo to be pending. He laments that the students are not well informed, but his hopes, building on what they were told more than three years ago, are high. It will bring a good future for Madagascar, Ilo strongly believes, because "it's a very big [infrastructural] project." Some more context on this "very big project" and Madagascar's potential involvement that already spurs emotions and local actions is necessary here.

The international headquarters of the SKA are located in the United Kingdom, and European countries contribute the vast majority of the project's funding (see Figure 4.3). South Africa is the only full member country of the SKA international consortium; the other eight African partner countries do not contribute to the budget directly. This uneven landscape emphasises South Africa's leadership role in the continent's science and technology initiatives that reproduce the unequal relations between metropolitan interests and the presumed "empty" hinterland on the continental scale. Thus, South Africa's hegemonic position in Africa, usually referred to in terms of its "exceptionalism" (Lazarus 2004), is a continuation of its global ambitions.

To understand the relationship between South Africa and its SKA African partner countries, we must take a step back to the early 2000s, when preparations were made in South Africa to harness Africa's potential for astrophysics. Radio telescopes that would be located in Africa – with its

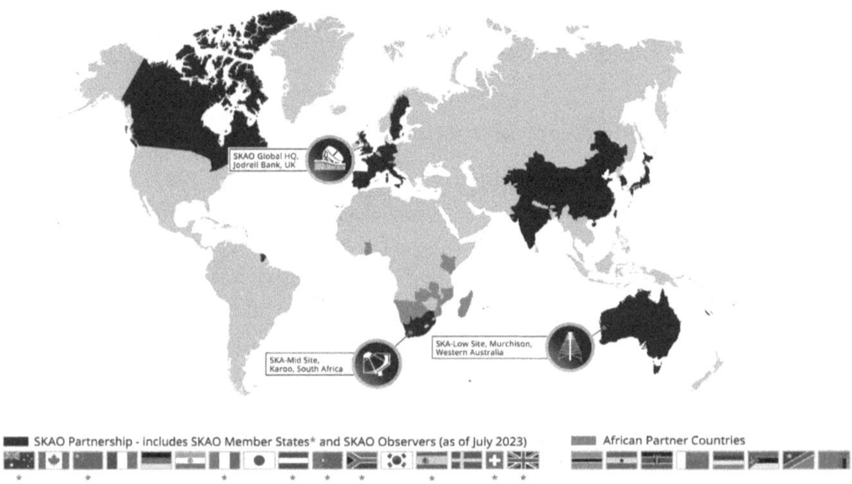

Figure 4.3 Map of participating countries in the SKA Organization. Credit: Square Kilometre Array Organisation.

considerable landmasses in the Southern Hemisphere – promised to yield scientific perspectives into outer space that are complementary to existing data and could thus complexify astrophysical observations. With a vision to profit from the astrophysical demand for African telescopes and to design a truly "African" bid for the SKA, designating the SKA as evolving into a pan-continental infrastructural system, South Africans sought existing large telecommunication facilities that could be transformed into telescope dishes. Subsequently, a South African delegation visited the countries where suitable candidates for renovation were identified. As a whole, South Africa assembled eight "African Partner Countries" that form the "African VLBI network" (AVN). The AVN is designed with the goal to expand the SKA across southern, western, and eastern African countries, effectively turning the African continent into a giant telescope to probe into the universe. South Africa's vision for the SKA was portrayed as an African vision; it proposed to meet the upscaling of astrophysical investigations by upscaling scientific infrastructures to continental levels. Although the actual implementation of the SKA's Phase 2, which would include converted satellite dishes in African partner countries, is indeterminate as of now, their anticipation alone has already sparked a number of local effects and impacted terrestrial imaginaries, which we exemplify with the case of Madagascar. Put differently, the poetics of astrophysical infrastructures precede their installation in Madagascar.

Figure 4.4 The disused telecommunication dish in Arivonimamo, Madagascar. Photograph taken by Hanna Nieber.

Like many current students, Ilo had heard about the SKA before he started his astrophysics studies. Indeed, the prospect of Madagascar hosting part of such a big infrastructure excited him to such an extent that it determined his choice of study. He and many other astrophysics students continue to express their desire for astrophysical infrastructure to be built in Madagascar and eagerly await news about the dish in Arivonimamo (Figure 4.4).

In Arivonimamo, approximately an hour's drive from Antananarivo, Madagascar's capital, the formerly state-owned telecommunications company built a telecommunication dish in 1972. When the first plans to include African partner countries took shape in South Africa, this disused dish became an attractive possible asset for the SKA. Its potential to be converted into a radio telescope put Madagascar on the map for South Africa's partner countries. Delegations from South Africa that visited the facility found the dish suitable, and advised Malagasy science professors to initiate the introduction of a programme for astrophysics at the state university. The astrophysics master's programme at the University of Antananarivo was inaugurated in 2014, from which Ilo later profited. Working towards inclusion in the SKA, the Malagasy Ministry of Higher Education and Scientific Research and the now private telecommunications company that owns the dish agreed on a 20-year lease in 2017 to enable conversion of the dish into a telescope. However, political shifts have resulted in unstable support for this project in Madagascar. Furthermore, the SKA itself withdrew support for the conversion, after a very similar project in Ghana proved more expensive than initially anticipated and did not yield the same scientific value as a newly built dish would have done. Nevertheless, holding on to Madagascar as one of its African partner countries, the SKA now plans to erect a new telescope in a more favourable location in Madagascar instead. This new dish has yet to materialise. In the meantime, Malagasy astronomy enthusiasts still harbour hopes to find funding for the Arivonimamo dish conversion. As an object that had attracted an interest from South Africa and that, for now, remains the only tangible materialisation of what could one day become an infrastructure for radio astrophysical research, the planned dish in Arivonimamo continues to nurture Malagasy astrophysicists' hopes and aspirations for a functional radio telescope on Malagasy grounds.

An island in the Indian Ocean, Madagascar often portrays itself as a "continent" in its own right (De Wit 2003; Randrianja and Ellis 2009). Overall sentiments towards the African mainland among the Malagasy are not particularly favourable, sometimes even condescending. Mamy, another astrophysics student and passionate about doing something good for her country, articulated this attitude: "When people would speak about Africa, I used to think that Africa is later than other continents [lagging behind]."[12] Mamy had spoken previously about her investment in social change and her volunteering work with children from impoverished backgrounds. She has a vibrant "let's-do-this" attitude to life, and, coming from her, the notion of Africa's inertness, its being late and lagging behind, is not a compliment.

But Mamy did not utter this sentence as a conclusive statement. Rather, she intended to provide context for the perception change that came about for her after getting involved in astronomy: "When I discovered astronomy, the first project I discovered was the SKA Project. And I was really amazed that Africa can host some big project like this!" While her surprise attests to the persistence of the derogatory stereotypes of Africa in Madagascar, it also marks the moment Mamy started to think about Africa in more positive terms. Here, astrophysical infrastructure serves as an indicator of development in Africa, one that is deemed desirable and becomes a potential locus of pride. The prospect of being infrastructurally linked to Africa allows Mamy to participate in this pride and in this large infrastructure's poetics. Another student and astronomy enthusiast, Ary, took this a step further. She said:

> I believe that the continent is developing significantly. [...] there are projects [...] that bring participant countries of the same continent together and for us, that's Africa [speaking about the SKA's AVN]. It's important that the countries of the continent maintain a certain [reliable] connection to make this concept work. So, Madagascar can develop more by getting closer to the countries of the African continent.[13]

Ary thinks that submitting to a predefined logic of belonging and improving Madagascar's relationship with Africa would be a strategic accomplishment. Taking the astronomical project as an example, she posits that individual projects may profit from continental cohesion, and indeed this is the discourse that the SKA promotes for the African partner countries. In the official SKA discourse, the engineering logic of assembling telescopes is directly linked to an ideational configuration of developing the African continent. In a recent podcast interview, the SKA Observatory Council Chairperson, Dr Catherine Cesarsky, said:

> SKA will continue what indeed MeerKAT has been doing extremely well. Which is, on the one hand, [to] attract young people in the country [South Africa] towards astronomy, perhaps, but I would say science and technology in general, and this is useful not just obviously for astronomy or for SKA, but it's useful for the country. Training young people to later work in many, many different avenues using very particular skills that can be learned through astronomy.
>
> It's also astronomy for development. SKA is doing it at a very, very high level. [...] they are giving scholarships, a large number of scholarships, many of them to Africans and not just from South Africa, at least 30 per cent have to be from other African countries.[14]

Astrophysics, here, is an attractive natural science, with transferable skills to other fields of science and technology. Like many other people, Cesarsky easily takes the discursive step from advancements in science and technology to development. In Madagascar, the discursive tropes that Cesarsky taps into are well-known and oft-repeated. However, while the SKA infrastructure is celebrated in the South African political discourse as enabling a grand idea of "Africa" to materialise, Ary maintains a more critical distance. For her, the SKA infrastructure brings development; "Africa" figures merely as a means for such development to also take place in Madagascar, taking advantage of the particularity of outer space's vastness:

> We study something that is very vast, that has a big scale, so many regions can become involved in the same project. In certain sciences, one can be interested in something very small and local, so one can think that local research suffices, but for astronomy that does not work.

Astronomical infrastructure, Ary makes clear, requires collaboration across local contexts because astrophysicists study "something that is very vast." She acknowledges that the infrastructural needs to study the vastness of outer space entail particularities to which Madagascar's terrestrial placement caters. Though marginal to and perhaps even distinguishable from "Africa," it sufficiently meets the geopolitical conditions of belonging to "Africa" to become encapsulated as an "African partner country" for the SKA. Ary thereby exemplifies how astronomy not only elicits excitement and motivates people to engage with science, technology, and big data, but also influences people's imaginaries of geopolitical transformations on earth. With the SKA, "Africa" becomes a meaningful point of reference that mediates students' relationship with outer space. With the anticipation of the infrastructure to arrive, this "Africa" elicits hope, desires, and potentials for pride in Madagascar – the changing disposition to "Africa" is an infrastructural effect that precedes the local materialisation of radio telescopes. Put differently, the desire to make Madagascar relevant to the study of the universe, spurred by the poetics surrounding the anticipation of astronomical infrastructure (Larkin 2013; see also Anand, Gupta, and Appel 2018), has turned "Africa" into a strategic epistemic object that facilitates hopes of inclusion, development, geopolitical transformations, and scientific access to outer space.

Conclusion

Space anthropology invites the methodological and epistemological challenge of attending concurrently to the vastness of outer space, the immanent histories of a small town in a remote province of South Africa, and the aspirational futures of astronomy students in Madagascar. These realms do not easily connect with each other. Existing and emerging infrastructures for outer space create tensions with local inhabitants in Carnarvon that Malagasy students of astrophysics do not consider; future infrastructures for outer space and their already existing poetics create hope and desires for astrophysics students in Madagascar that are not shared by the Carnarvon residents. Yet both are connected through an emerging infrastructure for research on outer space, the SKA, and it is through space anthropology's work across scales that an analysis of the SKA's relation to "Africa" as an epistemic object comes to the fore.

Discursively framed and politically supported as instantiation of the "African Renaissance," South Africa's investments in astrophysical infrastructures become subsumed in a logic of national modernity that leverages imaginaries of the continent's future – but radically de-emphasises the repercussions of such investments for local residents in Carnarvon. Here, "Africa" becomes a notion that supports South Africa in reconfiguring its territories and thinking beyond its national boundaries. It is this notion of "Africa" that local concerns need to be defended against. The local concerns at the site of a future telescope in Madagascar, on the other hand, are yet to emerge. Perhaps potentially difficult situations in Madagascar could be averted by taking the South African case into consideration, but perhaps each site comes with its own dynamics. While such a speculative story of a potential future could be told, in this chapter we are more interested in how the promise for astrophysical infrastructure in Madagascar is entangled with a re-evaluation of "Africa" and as such already has an impact in the present. The prevalent unease with which many Malagasy people regard their country's belonging to the African continent shifts for astronomy enthusiasts. Not least because it is "African," Madagascar was selected as a SKA "African partner country" which has already led to the inauguration of an astrophysics Master's programme at the University of Antananarivo. Promised investment in infrastructure and ongoing human capacity building, many young Malagasy believe, are ingredients for Madagascar's development – and this development comes through an "African" project for research on the universe. The SKA's framing of "Africa" may connect the case studies in Carnarvon and Madagascar; however, these cases in juxtaposition show how differently people relate to outer space research and its multi-scalar historical and contemporary repercussions, including notions of "Africa."

On a methodological level, we have worked through ethnographic and historical methods. We have used our ethnographic encounter with the SKA as an entry point to interrogate the stratified layers of politics and history behind Africa's move to space. The analysis of large-scale infrastructures for outer space research necessitates this, and in this chapter we have provided an example of how it can be achieved. This collaboration has allowed us to trace the emerging pan-continental astrophysical infrastructure operating as a "system" (Larkin 2013) and to take seriously how infrastructural connections forge engagement – albeit differently – with a connective epistemological object: Africa. In working collaboratively, we have pointed to the complexities that discursive delineations of outer space infrastructures elicit for intra- and inter-state relations as well as relations to larger – continental – structures.

Finally, we have engaged with the anthropology of infrastructure to interrogate how scalar relations between regions, from local to cosmic, are produced. Our fieldwork highlights two important aspects of the anthropology of space infrastructures. First, the hypervisibility of space infrastructure produces expectations of modernity. Second, the promise of astronomy infrastructure produces infrastructural effects, desires, fantasies, and hopes that precede the building and technical functioning of the telescopes. Taken together, these two aspects of astronomy infrastructure constitute the hinges by which outer space is connected to "Africa" in the grand narratives of the future and practical hopes for development.

Acknowledgements

This work is partly based on the research supported by the South African Research Chairs Initiative of the Department of Science and Technology and National Research Foundation of South Africa (Grant No 98765).

Notes

1 The term "astronomy" delineates the study of celestial objects and matter. "Astrophysics" is a branch of astronomy, theorising the structures of the universe by applying physical and chemical laws. In this chapter, we use them with flexibility to account for the language that is used by our interlocutors and in official documents.
2 This is similar to what Dimitris Dalakoglou (2010, 2012) termed infrastructural fetishism.
3 For more information visit www.skao.int/en [accessed on 21 November 2023]. The SKA is expected to pave the way to important scientific discoveries about the universe and to generate innovation spillovers to many sectors through a multiplier effect.
4 SABC news, 25 May 2012. See www.sabc.co.za/news/a/4b4dbc804b602a6c9be09f5b1193da06/Engineers,-scientists-to-benefit-from-SKA-project-20122505 [Accessed 25 August 2017].
5 See www.theregister.com/Print/2012/05/25/ska_shared/ [Accessed 24 July 2023].
6 See www.gov.za/president-jacob-zuma-visits-ska-project-karoo-celebrate-african-science-success-story.
7 See www.ska.ac.za/media-releases/meerkat-joins-the-ranks-of-the-worlds-great-scientific-instruments-through-its-first-light-image/.
8 Interview with a white farmer, Carnarvon, 14 March 2016.
9 Interview with a white farmer, Carnarvon, 25 March 2016.
10 Chinigò and Walker (2020), Walker and Chinigò (2018).
11 Interview with Ilo (pseudonym), 18 March 2022, translated by author HN.
12 Interview with Mamy (pseudonym), 23 February 2022.
13 Interview with Ary (pseudonym), 1 March 2022, translated by author HN.
14 Extract from the podcast *The Cosmic Savannah* [thecosmicsavannah.com], episode 59, 5 December 2022.

References

Anand, Nikhil. 2012. "Municipal Disconnect: on Abject Water and its Urban Infrastructures." *Ethnography* 13(4):487–509.

Anand, Nikhil, Akhil Gupta, and Hannah Appel. 2018. "Temporality, Politics, and the Promise of Infrastructure." In *The Promise of Infrastructure*, edited by Nikhil Anand, Akhil Gupta, Hannah Appel, 1–38. Durham: Duke University Press.

Archambault, Julie Soleil. 2012. "'Travelling while Sitting Down': Mobile Phones, Mobility and the Communication Landscape in Inhambane, Mozambique." *Africa* 82(3):393–412.

Barker, Joshua. 2005. "Engineers and Political Dreams: Indonesia in the Satellite Age." *Current Anthropology* 46(5): 703–27.

Barry, Andrew. 2001. *Political Machines*. London: Athlone.

Beinart, William, and Saul Dubow. 2021. *The Scientific Imagination in South Africa: 1700 to the Present*. Cambridge: Cambridge University Press.

Belay, Yeshiwas Degu, Emanuele Fantini, and Iginio Gagliardone. 2020. "The Grand Ethiopian Renaissance Dam: Media Narratives and State Building." *Afriche e Orienti* 2: 27–48.

Chinigò, Davide. 2019. "From the 'Merino Revolution' to the 'Astronomy Revolution': Land Alienation and Identity in Carnarvon, South Africa." *Journal of Southern African Studies* 45(4): 749–766.

Chinigò, Davide and Cherryl Walker. 2020. "Science, Astronomy, and Sacrifice Zones: Development Trade-offs, and the Square Kilometre Array (SKA) Radio Telescope Project in South Africa." Social Dynamics 46: 391–413.

Dalakoglou, Dimitris. 2010. "The Road: An Ethnography of the Albanian–Greek Cross-Border Motorway." *American Ethnologist* 37(1):132–149.

Dalakoglou, Dimitris. 2012. "'The Road from Capitalism to Capitalism': Infrastructures of (Post)Socialism in Albania." *Mobilities* 7(4): 571–86.

De Wit, Maarten J. 2003. "Madagascar: Heads It's a Continent, Tails It's an Island." *Annual Review of Earth and Planetary Sciences* 31: 213–48.

Dubow, Saul. 2019. "200 Years of Astronomy in South Africa: From the Royal Observatory to the 'Big Bang' of the Square Kilometre Array." *Journal of Southern African Studies* 45(4): 663–87.

Edwards, Paul N. 2003. "Infrastructure and Modernity: Force, Time, and Social Organization in the History of Sociotechnical Systems." In *Modernity and Technology*, edited by Thomas J. Misa, Philip Brey, and Andrew Feenberg, 185–225. Cambridge, MA: MIT Press.

Ferguson, James. 1999. *Expectations of Modernity: Myths and Meanings of Urban Life on the Zambian Copperbelt*. Berkeley: University of California Press.

Grilli, Matteo, and Frank Gerits (eds). 2020. *Visions of African Unity: New Perspectives on the History of Pan-Africanism and African Unification Projects*. Cham: Palgrave Macmillan.

Larkin, Brian. 2013. "The Politics and Poetics of Infrastructure." *Annual Review of Anthropology* 42: 327–343.

Lazarus, Neil. 2004. "The South African Ideology: The Myth of Exceptionalism, the Idea of Renaissance." *The South Atlantic Quarterly* 103(4): 607–628.

Mavimbela, Vusi. 1998. "The African Renaissance: A Workable Dream." In *South Africa and Africa: Reflections on the African Renaissance*, edited by G. Pere, A.

Nieuwkerk, and K. Lambrechts, 29–34. South Africa: Foundation for Global Dialogue, FGD Occasional Paper No. 17.

Miescher, Stephan F., and Dzodzi Tsikata. 2009. "Hydro-Power and the Promise of Modernity and Development in Ghana: Comparing the Akosombo and Bui Dam Projects." *Ghana Studies* 12(1), 15–53.

Mitchell, Timothy. 2002. *Rule of Experts: Egypt, Techno-Politics, Modernity.* Berkeley, CA: University of California Press.

Mitchell, Timothy. 2011. *Carbon Democracy: Political Power in the Age of Oil.* London: Verso.

Mrázek, Rudolf. 2002. *Engineers of Happy Land: Technology and Nationalism in a Colony.* Princeton, NJ: Princeton University Press.

Nieber, Hanna. 2024. "Universality as Horizon: Aspirations and Geometries of Astrophysics in Africa." *Zeitschrift für Ethnologie | Journal of Social and Cultural Anthropology* 149 (1): 53–70.

Patel, Prina. 2016. "Africa's MeerKAT 'First Light' Images Have Blown All Expectations." The Conversation, 28 September. https://theconversation.com/africas-meerkat-first-light-images-have-blown-all-expectations-65246

Pović, Mirjana, Michael Backes, Paul Baki, David Baratoux, Solomon Belay Tessema, Zouhair Benkhaldoun, Michael Bode et al. 2018. "Development in Astronomy and Space Science in Africa: Comment." *Nature Astronomy* 2: 507–10.

Randrianja, Solofo, and Stephen Ellis. 2009. *Madagascar: A Short History.* Chicago, IL: University of Chicago Press.

Redfield, Peter. 2000. *Space in the Tropics. From Convicts to Rockets in French Guiana.* Berkeley, Los Angeles, London: University of California Press.

Republic of South Africa (ROSA). 2020. "Declaration of Certain Properties Situated in the Northern Cape Province the MeerKAT National Park." Proclamation No. 15 of 2020 under the National Environmental Management: Protected Areas Act, 2003. Government Gazette, no. 43145. 27 March. Cape Town.

SARAO. 2017. *SKA and Agri SA Partners for the Benefit of Local Communities.* Last modified 2 February 2017. www.ska.ac.za/media-releases/ska-and-agri-sa-partners-for-the-benefit-of-local-communities/

Star, Susan L. 1999. "The Ethnography of Infrastructure." *American Behavioral Scientist* 43(3):377–91.

Star, Susan L., and K. Ruhleder. 1996. "Steps Toward an Ecology of Infrastructure: Design and Access for Large Information Spaces." *Information Systems Research* 7(1): 111–134.

Von Schnitzler, Antina. 2008. "Citizenship Prepaid: Water, Calculability, and Techno-Politics in South Africa." *Journal of Southern African Studies* 34(4): 899–917.

Walker, Cherryl, and Davide Chinigò. 2018. "Disassembling the Square Kilometre Array (SKA): Astronomy and Development in South Africa." *Third World Quarterly* 39(1): 1979–1997.

Walker, Cherryl, Davide Chinigò, and Saul Dubow. 2019. "Introduction. Karoo Futures: Astronomy in Place and Space." *Journal of Southern African Studies* 45(4): 627–639.

5 Museums, meteorites, and portals

Tracing the imperial logics of trans-planetary resource extractivism

Alana Osbourne

By way of introduction

'This is the oldest object you will ever touch...'

It was a drizzly Sunday morning. My sister, her kids, and I decided to spend it sightseeing in Greenwich, visiting the Royal Museums that were once foundational institutions of the British Empire. My family had been there before, but I hadn't, and in my excitement I made a series of bookings: a ticket to stand on the GMT line and seats for a planetarium show at the Royal Observatory. We started by heading up Greenwich Park, marvelling at the wet, panoramic view of North London before walking the path that leads to the observatory. Signs indicated that the planetarium shows took place on the underground level of the small, rounded building, and my 10-year-old niece and 12-year-old nephew raced down ahead in the hopes of securing us good seats.

When I reached the bottom of the building's winding staircase, I found my niece standing in front of a large chunk of grey-brown rock, the size of a small dog, placed on a sleek display table. It was the only object exhibited in this small circular gallery and positioned by the staircase it lay en route to the planetarium. Looking at the rock, my niece raised her eyebrows and scrunched up her mouth, an early teenage expression I had come to understand as meaning 'unimpressed.' She stepped back from the rock, took in the caption printed on a tall white panel, and read it aloud: 'This is the oldest object you will ever touch.' She then read the short explanatory text stating that the rock, composed of iron and nickel, was a piece of a larger meteorite known as the Gibeon that had hit the Earth in prehistoric times, landing in what is now Namibia. She looked at the rock again, her head slightly tilted, pensive. She touched it, briefly intrigued, then walked away. Her older brother, who had been standing behind her, stepped forward and gently glided his hand along the rock's cool surface, whispering a soft and engrossed 'waow.' Not only was this rock the oldest

DOI: 10.4324/9781003437956-5

object he had ever touched, it was also alien matter. It had come from 'outer space,' a fact he later recalled with awe. We stood a little longer by the stone. He took a picture of it on his phone, poked his index into one of the meteorite's small crevasses, as if measuring it for size, then shuffled on towards the Observatory's planetarium. A show about the Moon and the possibilities of mining lunar ice was about to start.

Throughout this chapter, I rely on and return to this family day trip, which took place in the summer of 2023, to think about the relations between museums, imperialism, rocks and outer spaces. Much has been written about the supporting roles that museums play in making and extolling empires (Bennett 1995; Azoulay 2019; Hicks 2020), with the core of this literature reminding us that museums are public interfaces through which colonial aspirations are promoted or justified (Bennett et al. 2017). Elsewhere in this volume, Makar Tereshin and Denis Sivkov (Chapter 6) extend this proposition by showing us how museums that chronicle the past and future of space travel are also loaded with politically charged discourses that include narratives of discovery, subjugation, and exploitation. At the Royal Greenwich Observatory, a site with deep historical roots to the British Empire, new connections between museums and empire are made apparent through planetarium shows that pitch outer space as a vast expanse to be explored, conquered, and from which to extract wealth. While the notion of extractivism is most apparent in the planetarium's segment on Moon mining, in which the speculative quarrying of alien grounds is detailed through 3D animation, in this chapter I focus on the fragment of rock exhibited at the bottom of the observatory's staircase. Noticing that the presentation and the materiality of the Gibeon meteorite support the narrative of resource mining later expounded in the planetarium shows, I contend that the observatory's simple, sober display of rock is imbricated in complex imperial histories and futurities.

Mining, and the related science of geology, are central to colonial projects (Yusoff 2018; Lester and Pico 2021a, 2021b; Liboiron 2021); geological rooms built in the early twentieth century in museums across Europe and the US attest to this. In these displays, ores and minerals served to entice business investments and public backing for quarrying enterprises (Gelsthorpe 2021; Hearth and Robbins 2022) and created a proximity between museum audiences and colonised lands. Filling the gap of geographical distance with narratives of earthly riches, mineral cabinets offered a material sense of familiarity with arrogated spaces (Osbourne, forthcoming). Throughout this chapter, I show how similar designs underpin meteorite displays. Focusing on the Gibeon fragment, I outline the ways in which alien rocks are used to support discourses about prospective extractive missions in outer spaces. While these projects are still hypothetical, it is important to note that the composition of iron and nickel-rich meteorites, such as the Gibeon, is analogous to rocks currently quarried in formerly colonised terrains and used – amongst other things – for the development of digital technologies. As such, just like earthly rocks, meteorites are often valued according to their potential function as resources, thereby extending the logics of extractivism to encompass matter beyond our orbit.

Before I unpack the connections between meteorites, museology, and extractivism, I start by outlining how scholars have proposed to engage ethnographically with spaces beyond our planet. Drawing from this review, and adding to the conversations compiled in this publication, I emphasise that outer spaces are ethnographically researchable on Earth (Messeri 2016), and through inorganic material. In other words, via the observatory, and with reference to the Gibeon fragment on display, this chapter offers a *grounded* ethnography of outer spaces and develops propositions to think about the politically charged entanglements through which related geologic and extractive processes unfold on and beyond the globe. To this end, I rely on the concept of 'planetary portals' that Casper Laing Ebbensgaard, Kerry Holden, and Kathryn Yusoff (2022) use to highlight interconnected geographies of extractivism that operate across temporal and spatial divides. The portal, they argue, is an analogy that enables us to visualise linkages between extracted zones, inorganic matter, and localities of wealth accumulation as they shift – yet continue to operate – through time. Extending their analytic, I suggest that placed in the Greenwich Observatory, the Gibeon meteorite is part of a 'trans-planetary' portal that spreads out further in space and time by deploying the same 'geo-logics' (Yusoff 2024, 28) of extractivism onto alien matter.

Re-fielded terrains and trans-planetary portals

Thinking about outer spaces, many scholars have highlighted the difficulties posed by studying a spatial domain that remains, for most of us, physically unattainable. Acknowledging this tension and drawing from a breadth of work that critically assesses how we produce our fields of enquiry (see Haraway 1988; Berry et al. 2017), Valerie Olson (2023) writes against the notion that to research extraterrestriality one needs to be physically located above the Earth's atmosphere. Rather, referencing emerging literature in the social sciences of outer spaces, she highlights how researchers have 're-fielded' the terrain of their investigations by analysing the processes, infrastructures, and networks that cut across and connect a range of social practices and geographical locations, including the cosmos. Drawing on the works of archaeologists who analyse temporally and geographically remote launch sites and space debris (Gorman 2005a, 2005b; Gorman and O'Leary 2013), and Indigenous scholars who foreground the cosmopolitical dimensions of 'the field' (Atalay et al. 2021; Lempert 2021), Olson emphasises how objects and processes that structure human sociality can be both firmly anchored on Earth and tied to outer spaces (2023). Re-fielding our ethnographic grounds in this fashion implies, in part, de-centring the importance of a shared geographical experience and demands that we include outer space things, such as the sun, moon, and planets, in our analysis of social experience (Hobart 2019; Shorter and TallBear 2021). Such an approach makes room for centring the porousness of the cartographies we work with and unbinds the categories we often reason with. It allows us to scramble the here and there of topographical and temporal separability.

Consonant with Olson's re-fielding, Anne Beaulieu's work in STS (2010) suggests a departure from ethnographic practices that rely squarely on shared location to think through moments in which researchers are 'co-present' with their terrain. 'Conceptually,' she writes, 'co-presence foregrounds the relationship between self and other and interaction that *achieves* presence in a setting' (2010, 457). Using telephone conversations to illustrate her proposition, Beaulieu shows how co-presence hinges on interactions that require specific infrastructures to be sustained. In the case of phone communications, these arrangements include the tones, signals, and technological networks that enable and punctuate social relationalities. Transposed to the social studies of outer space, Beaulieu's insights enable Victor Buchli (2020) to think through the multiplicity of locations – Earth-bound and extraterrestrial – that co-construct life on the ISS positioned in Lower Earth Orbit. Focusing on the synchronised calendars, routines, and work practices through which actors based in countries across the globe shape daily life on the ISS, Buchli shows us how studying extraterrestriality can contend with a plural spatiality in ways that do not necessitate the ethnographer to be physically located in outer space. The forms of co-presence that constitute life on and below the ISS, along with the infrastructures through which they are made possible (see also Bichsel 2020), scramble the separability between terrestrial and cosmic realms (Battaglia et al. 2015; Valentine 2016). What transpires across these co-presential and re-fielded ethnographies is an invitation to collapse geographical scales we often rely on, and to research the various forms of relationality that connect the body to the expanse of the universe (see Harjo 2019).

In this text, I bounce off of Olson's and Buchli's invitations to rethink the locations of outer space ethnographies and offer the museum as a place through which cosmic intimacies and 'more-than-Terran' (Olson 2023) geographic entanglements are researchable. As I unpack in this chapter, trips to Greenwich, and more specifically, encounters with the Gibeon meteorite, exemplify a relationality with celestial bodies and outer-space dynamics that invoke multiple locations. Because of this, I understand this meteorite fragment as opening '*portals* into new spatial and theoretical configurations of structure, relations, subjectivity, and time' (Olson 2023, 33, my emphasis). I stress the word 'portals' in Valerie Olson's quote because I believe this is a particularly generative term to think through connectivity without proximity, especially when relations between earthly and terrestrial spaces are mediated through inorganic material. Thinking with portals can reveal how outer space and imperialism are entwined, and how poking the crevasses of a meteorite on a Sunday morning might intimately connect museum-goers to past and future imperial projects of extractivism, on and beyond our planet.

To understand how a piece of dull-brown rock fallen from the sky can be used to herald the lucrative importance of space exploration, and to see how this connects to earthly mining and imperial continuities, I draw on the work of the 'planetary portals collective' whose members Kathryn Yusoff, Kerry Holden, Casper Laing Ebbensgaard and Michael Salu deploy the 'portal' as an analytic (Yusoff, Holden, and Laing Ebbensgaard, 2022). Outlining transcontinental links between the Cecil Rhodes archives in Oxford, mines in South Africa, and bricks in London, they use this concept 'to map the interconnected geographies and afterlives of colonial infrastructures; constituting spatial imaginaries that were deployed as blueprints in the emergence and maintenance of extractive planetary futures' (2022). The portal reveals the imperial processes of resource extraction by connecting the multiple geographies through which mined matter flows and is transformed. Mobilising this concept, I suggest that the Gibeon meteorite, as exhibited in the museum, extends the portal into a *trans*-planetary device connecting locations far above and far below terrestrial grounds, and through both past and future temporalities. Relying on the image of a gateway to think about the Gibeon fragment enables two conceptual moves. First, it allows us to consider inorganic material, such as the meteorite, as linked to the different spaces and times that are chronicled by the museum's narrative. Second, it allows us to trace shifts in the meanings ascribed to meteorites as they circulate through seemingly disconnected places and times to sustain networks of resource extraction. I develop each of these propositions in the following sections of this chapter.

The imperial museum matrix

The portal enables us to imagine the meteorite as an entry point into the complex histories and temporalities catalogued by the Royal Observatory, the location of its display. As I unpack below, the Greenwich Royal Observatory was an instrument of territorial conquest for the British

Empire, and it constitutes part of a colonial frame that stretches horizontally across the globe. Via the observatory's focus on the skies, this imperial grid also spreads vertically toward outer spaces yet to be explored. Although the Royal Greenwich museums develop a (tentatively) self-reflexive narrative about the role that their institutions played in the brutal processes of empire making and maintaining, the structure of the buildings and the romanticised stories of discoveries and expansion that they host are still imbued with pride for the nation's history of conquest. This means that the location, design, and curation of the observatory into a public-oriented learning space is laden with stories of exploration, innovation, and domination. Hosted within this matrix, the Gibeon fragment is enrolled into this narrative and becomes an important actor in a plot centred on territorial expansion.

The Royal Observatory was founded in 1675 and is located in east London, on the same grounds as the National Maritime Museum, the Queen's House, and the Old Royal Naval College. The four institutions collectively make up the Royal Museums Greenwich (RMG) and conjointly demonstrate the importance of the seas, ships, time, and stars in British history.[1] According to descriptions on the RMG website, the observatory was initially built to 'harness astronomy' and to offer a cartography of the sky that could 'support expanding British trade and naval power.'[2] The naval power in question was, of course, essential for the creation and enlargement of the British Empire. Dominating the seas enabled Britain to become, by the end of the eighteenth century, a main partaker in the transatlantic slave trade and plantation economies. Astronomy, nautical engineering, and time-gauging devices, developed in part at Greenwich, were among the crucial technological advancements required to support this brutal expansion into – and control over – far-flung territories and people.

This rapport between clocks, the skies, and the oceans is one that is carried through today not only by the observatory's close geographic location to the Maritime Museum but in the parallels between the collections held in each building. While the Maritime Museum emphasises the engineering feats of late sixteenth-century marine ships, the Royal Observatory references multiple outer space vessels. Together, these exhibits draw attention to the mobility technologies that enable (extra)territorial explorations and are required to navigate uncharted grounds. While the collections connect oceanic and celestial depths with travel and engineering, they are buttressed by the Royal Museums' principal attraction: the Prime Meridian. Cutting the globe from North to South, meridians serve as reference lines for astronomical observations and, by dividing the West from the East, they enable the measure of terrestrial distance. In the nineteenth century, the Prime Meridian also became the point of reference for homogenising global clock time – otherwise known as Greenwich Mean

Time. As Rasheeda Phillips notes in a beautiful sequence of essays (2018, 2019, 2021), Mean Time was more than a temporal index: it was a colonial tool of control, and oppression. Through homogenised time keeping, imperial powers coordinated the logistical chains through which matter could be extracted from colonised grounds and imposed a standard through which to measure and regiment the labour of subjugated populations.

While the stars served to construct the devices and logics that synchronised time and enabled the meridian dicing of the globe, in a reverse movement, Greenwich Mean Time now organises life in the cosmos. In his work on the ISS, Buchli spotlights the imperial underpinnings of the relationship between time and the stars in a passage worth quoting at length. As he explains;

> The ISS experiences 15 or 16 sunrises and sunsets in a day producing a radically distinctive time/space from that on Earth despite the coordination of time on the ISS with GMT. The GMT standard was established by the British Empire to regulate imperial time/space, and its use extraterrestrially represents an echo and refiguring of that imperial order as a compromise between two other competing orders on the ISS, the Russian (Soviet) and American. The two realms – the extraterrestrial ISS and the terrestrial realms of its mission controls – are kept in tune with each other through the extension of this vestige of British imperial seafaring chronometry.
>
> (Buchli 2020, 25)

Besides clarifying enduring connections between outer spaces, seas, and time, Buchli's account of synchronised time highlights how Britain's imperial grid extends horizontally across oceans and vertically beyond our orbit, with the observatory and Prime Meridian at its centre.

If the wider museum context celebrates nautical engineering and time-keeping feats that enabled this imperial network, the planetarium shows played inside the observatory speculatively project it into the future. For our Sunday family visit to Greenwich, I had pre-booked tickets to a show that focused on the Earth's moon and, after patting the Gibeon meteorite fragment, we settled at the back of the planetarium to watch it. The show was impressive. It featured a well-crafted combination of archival footage depicting Moon landings, 3D animated segments on future lunar ventures, and recent photographs collected by robots that detailed the celestial body's dark side. In these images, the abysmal crevasses of the Moon's Southern Pole – in which frozen waters lay – are revealed. Over the images, the voice of the in-house narrator described how these discoveries had re-ignited a race to mine lunar ice. 'Artemis,' she explained, 'is the name of the next crewed Moon landing scheduled for late 2024 and it will launch the start of a decade-long scheme by the end of which lunar ice could be mined.' Conveying awe for these prospects, the narrator's tone added to the sense of marvel sustained by the immersive medium of surround-sound and the enveloping width of the dark, curved screen.

The show ended on this future-oriented note, and we walked away with a sense of the immense potentials of outer space. As our little group made its way out, blinking in the now dazzling brightness of the dome, my niece asked me if lunar ice was different from 'our ice,' and why we would need it. Unsure of my answer, I suggested we walk back into the planetarium to ask the narrator. Adjusting the volume on her mic to a lower setting and shutting down the projection system, she explained that lunar ice, just like ice found on earth, contains hydrogen, which could be used to fuel rockets. Landing on the Moon from our planet, rockets could load up on hydrogen-based propellants and restock water, before setting off to explore further frontiers. 'We are close to being able to leapfrog ahead, and maybe mine more valuable things that were too far away before,' she explained. Undisturbed by the narrator's focus on the projected uses and values of ice for further extractive missions, my nephew answered 'Leapfrog?! Coooool.' My niece mimed a hopping movement with her hand, conjuring the image of a toad as we walked out of the theatre.

The scientific missions and projections to excavate alien grounds, such as those sketched by the planetarium show, echo the narratives of exploration and conquest that shape the wider RMG institutions. When coupled with the Maritime Museums' focus on the vessels and technological developments that enabled naval mobility, travel beyond our orbit becomes a continuation of British imperialism across oceans. Critical scholars have already outlined the colonial logics of many off-world activities, and they have unearthed the rhetorical resonances between space exploration discourses and historical narratives of colonial expansion (Haskins 2018, Hobart 2019, McKinson 2020). For instance, unpacking a lexicon that un-reflexively deploys terms like 'New World,' 'frontier,' 'unexplored territory,' or 'terra nullius' to describe the cosmos, these authors have shown how the semantics of outer space travel revive Euro-American settler myths of rightful appropriation (Smiles 2020). This repertoire of terms used during the creation and expansion of Euro-American empires has now spilled into the vocabularies of space agencies and private investors, trickling all the way to planetarium shows, to reveal 'the continuing logics of settler colonialism, as well as questions of its future trajectories' (Smiles 2020, 1).

Departing from a morning spent at the Observatory with my family I have, thus far, outlined the historical and geographical processes of imperialism that are indexed in the RMG institutions. These imperial webs connect Greenwich to conquered territories by way of time-gauging devices and sea vessels. And, as the planetarium shows clearly articulate, such webs already extend into the cosmos, where projects to dig for ice anchor space ventures in a tradition of Euro-American colonial practices. In what follows, I focus more explicitly on the materiality of these imperial imaginings. Zooming in on the Gibeon's display, I trace how, by virtue of coming from outer space and containing extractible ores, the meteorite fragment features within this matrix. Reactivating the portal analytic, I detail the various meanings ascribed to the meteorite's substance as it moves from the cosmos and enters the museum. These shifts in meaning, I argue, are important. They allow museum-goers to perceive the rock not only as alien matter, but as an extractible resource – much like earthly minerals and ores – thereby strengthening the planetarium's narrative.

From alien rock to geological sample

Narrowing in on the curatorial choices used in the Gibeon's presentation, in this latter part of the chapter I suggest that via precise display choices, a sense of intimacy between visitors and alien rock can emerge and sustain dreams of outer space extraction touted in the planetarium. Staging here is important. As Anna Tsing emphasises in an essay that recounts how a Canadian prospecting company swindled investors by fabricating its claim to have struck gold in Indonesia, mining ventures 'must exaggerate the possibilities of their mineral finds in order to attract investors so that they might, at some point, find something' (2000, 118). Dramatisations and presentation, she argues, are key to nourishing extractive dreams, and I suggest here that the museal display of the Gibeon operates as a staging of minerals in support of extractive prospects in the cosmos. As I detail below, by offering visitors the opportunity to create an individual sensory connection with the Gibeon, and by relating the rock's trajectory, intimate channels can emerge between visitors and inanimate materials (Critchley 2008). This curated closeness is important: it allows the Gibeon fragment's meaning to vacillate (see Jeevendrampillai and Fortais, Chapter 9 this volume) between revered extraterrestrial stone and a geological specimen whose composition justifies quarrying beyond our orbit.

Such shifts in matter are central to processes of resource extraction. When withdrawing materials from the environment for human use, both inorganic substances and organic bodies are transformed from raw elements into wealth. By tracing these shifts, the portal analytic proposes to focus on the material dimensions of imperial grids. As the members of the collective explain, 'the portal is a way to see imperial imaginings as not a projection over space but a transformation of its temporal-material dimensions." What is important here is the "transformation of matter, changing *states of matter* and making states of political, economic, and ecological change' (Yusoff, Holden, and Laing Ebbensgaard, 2022; original emphasis). Unlike the transformations of matter recorded by the 'planetary portals collective,' in which South African diamonds become wealth and eventually the bricks of London's real estate, the materiality of the Gibeon does not (yet) shift. But its signification changes, and consequently the processes it indexes are extended. As Christopher Tilley tells us, 'stones' such as the Gibeon fragment 'always have meanings and relationships extending beyond themselves. [...] They are always more than themselves, in a process of becoming rather than a static state of being' (Tilley 2004, 222). Indeed, as I relate hereafter, the Gibeon fragment vacillates from alien matter and museal material culture, to being a geological specimen comprised of metals.

Because they come from outer spaces and hold important clues for a better understanding of our solar system, meteorites such as the Gibeon generate an amazement that radiates beyond communities of scientists and space enthusiasts. Amongst other things, this symbolic value ensures that they are treated with awe and respect within museal spaces. Yet they also contain ores such as iron and nickel, and high concentrations of valuable elements found on Earth, such as cobalt, that are crucial to the construction of batteries and electrical circuits, themselves vital to the aerospace industry. This chemical composition fuels dreams of mining in outer space (see Olson 2012). It is this vacillation between categorisations – from extraterrestrial messenger to mineable matter – that the Gibeon's exhibit operationalises in, first, conveying marvel for the rock and, then, enfolding it into future extractive dreams via the planetarium show.

The Royal Observatory's Gibeon display is by no means ostentatious, yet it manages to communicate wonder to the viewer. On the white panel, just underneath the text that reads 'this is the oldest object you will ever touch,' the specimen's age is detailed with the following caption: 'It is about 4.5 billion (that's 4,500,000,000) years old.' The repetition of the meteorite's age with numerous zeros in parenthesis casually serves to highlight the enormity of a quantity poorly encapsulated by the word 'billion.' As Dudley (2011) notes, citing factual criteria such as a displayed object's age or rarity, changes and engages the viewers' gaze, as exemplified by the shift in my niece's attitude. Amazement at the temporal immensity condensed in the rock echoes the fascination that meteorites have exerted throughout history. Already in the 1900s, anthropological texts concerned with cosmic folklore posited that 'there has probably never been a day when there was not being carried on, somewhere upon the globe, the worship of a "sky stone"' (Farrington 1900, 199). Considered ancient witnesses of outer spaces, and imbued with spiritual significance, they were, and still are, believed to hold (geological) secrets to the universe.

Echoing this, the numbers stated in parenthesis in the Gibeon fragment's caption rope museum-goers into this long-held sense of admiration.

As Daniel Sage (2014) notes, the wonderment inspired by extra-terrestrial ventures and materiality – or what he terms the 'cosmic sub-lime' – holds significant political sway. As he argues, it is partly through the awe inspired by NASA's exploits that American exceptionalism is reinforced. Similarly, it is through the reverence inspired by the meteorite that extractive futures are pitched to museum-goers. Yet, to articulate these potentialities, meteorites must be simultaneously celebrated and desacralised. While the display confirms that the Gibeon is the subject of wonder (it also features prominently on the museum's website as one of the key attractions), in contrast to the precious earth minerals displayed in curiosity cabinets in natural history museums across Europe and the US (Hearth and Robbins 2022), it is not encased in glass. And unlike the small but prized chunk of the Moon on show at the Science Museum in London, the Gibeon fragment is not embellished by a direct spotlight detaching it from a dark backdrop. It just lays bare, at the bottom of the staircase, available to the touch. Rather than remaining alien and unreachable, through touch the meteorite is made familiar and, by proxy, so are extra-terrestrial grounds. Through this curatorial choice, the meteorite fragment becomes a precious, awe-inspiring – *yet accessible* – object.

Within museum spaces, invitations to experience displayed objects through touch are rare. As Fiona Candlin reminds us (2008), Western sensory cultures are largely ocular-centric, with vision often understood as the dominant sense (Classen 2005). This emphasis on the visual has shaped museal practices – from curiosity cabinets to educational videos – and access to natural history collections is still mostly organised through ocular relationalities, in which materials are exhibited for the benefit of the visitor's eyes. Traditionally, keeping objects away from touch maintained their physical and symbolic integrity, so that tactile experiences were preserved for experts and for the privileged few (Chatterjee 2008). More recently however, museum scholars have explored the role of touch in collections, emphasising the importance of this sensory medium for the comprehension of displayed materials. Exploring how touch can be used in cultural institutions to facilitate understanding and learning, Elizabeth Pye posits that it enables an intimate gauging of the unmeasurable (2008), giving a sense of unfathomable units, such as, for instance, the Gibeon's age.

Intimacy isn't only kindled through touch, but also bolstered by the description of the Gibeon fragment found engraved in a metal plaque next to it. The narrative choice of the caption offers biographical elements that contribute to create a sense of warmth and proximity between inorganic matter and human visitors. Unlike the dispassionate descriptions of minerals and ores exhibited in natural history galleries (Knell 2000; Daston 2000), which often give little more than the sample's date, place of collection, and chemical composition (Fortey 2000), the description of the Gibeon constitutes a mini 'object biography' (Gosden and Marshall 1999) that retraces to meteorite's journey through time and space. It reads:

> This is part of the Gibeon meteorite that is believed to have hit the Earth in prehistoric times. Gibeon is an iron and nickel meteorite and was found as hundreds of fragments in Namibia, Africa. The iron and nickel form intricate patterns inside the meteorite. These patterns tell us that the metals cooled slowly over thousands of years as the meteorite travelled through space.

As Adam Drazin notes, providing object biographies is not a neutral act; rather it purposefully 'casts objects in an intimate light' (2020, 64). In the case of the Gibeon meteorite, it fosters closeness between museum-goers and rock, rendering it less alien. Indeed, offering this narrative, the museum curators – perhaps inadvertently – propose that the Gibeon be apprehended as a recognisable rock, a geological specimen composed of 'iron and nickel.' It presents the meteorite as an 'immutable mobile' (Latour 1990, 44–47, quoted in Chalk 2012), as an object that came to Earth already carrying geological information, rather than as having had chemical interests ascribed onto it by scientists trained within the enduring framework of extractive geo-logics. This presentation of the meteorite as a known scientific substance allows museum-goers to interact with meteorites as they would relate with earthly material. By making the alien familiar, extraterrestrial and earthly rocks become analogous, collapsing otherwise incommensurable scales of separability, and incorporating meteorites into a geological fold.

Zooming in on the Gibeon's display reveals how the meteorite fragment bolsters the Greenwich Observatory's narrative of space exploitation and extractivism. Despite the sobriety of the meteorite's presentation, the curatorial choices that frame it participate in kindling closeness between visitors and extraterrestrial matter. As I have detailed, the invitation to touch it, coupled with an affective and biographical description of its materiality, desacralise the meteorite and offer intimate channels for museum-goers to interact with alien matter, thereby making the plans detailed in the planetarium shows more tangible. These plans, which include mining the moon for lunar ice and leapfrogging to further extraterrestrial territories, are themselves a continuation of past projects to conquer and extract wealth from earthly lands. As the observatory's proximity to – and curatorial entanglements with – the other RGM institutions reveals, from the onset, Britain's imperial dreams of expansion across the globe also engulf the stars.

Conclusion

On the sluggish, traffic-jammed way back from our day in Greenwich, I asked my niece and nephew what they had thought of the planetarium show. After praising the quality of the 3D animated segments, my nephew brought up the meteorite fragment. 'It was kinda cool touching alien stuff,' he said. Then shifting his tone to adopt the voice of a cartoon-villain, his hand turned upwards and fingers flexed liked claws, he snickered 'and to think of all the power we will have once the moons, meteorites, and asteroids are ours, muahahaha.' Tracing the changing significance of the Gibeon, as well as the multiple historical processes that give it meaning, this chapter has detailed how earthbound displays of rocks are tied to complex patterns in which museums, outer space, and imperialism are connected through the logics of extractivism. In so doing, it attempted to unpack the association between the meteorite, the planetarium shows, and the imperial dreams of outer orbit extractivism that my nephew's evil emperor impersonation jokingly encapsulated.

To connect the meteorite to the history of extractivism and territorial conquest that cuts across the museums, and to trace how these projects imply a transformation of raw matter into wealth, I have drawn on the concept of 'the portal' (Laing Ebbensgaard et al. 2022). Departing from the Gibeon fragment, I have shown how the portal, initially deployed to think through Cecil Rhodes' imaginings of the African continent as a source of removable wealth, can similarly be deployed to unpack the Royal Observatory's narrativisation of outer spaces as repositories of mineable matter. This analytical move highlights how our Earth's bedrock is materially and ideologically entwined with outer spaces. Operating a 'trans-planetary' extension of the portal, this chapter has built on the work of scholars who explore 'human-but-not-only' (de la Cadena 2014) processes of extractivism. In so doing, it confirms that more-than-terran geographies and grounded ethnographies of outer space can reveal how the violent and extractive imperial projects that have re-mapped the globe also chart our relationships to the cosmos.

Acknowledgements

The author would like to thank Thijs Jeursen and Francesco Colona for their feedback on an earlier version of this chapter.

Notes

1 www.gov.uk/government/organisations/royal-museums-greenwich
2 www.greenwichworldheritage.org/maritime-greenwich/our-history

References

Atalay, Sonya, William Lempert, David Delgado Shorter, and Kim TallBear. 2021. "Indigenous Studies Working Group Statement." *American Indian Culture and Research Journal* 45(1): 9–18.

Azoulay, Ariella Aïsha. 2019. *Potential History: Unlearning Imperialism.* London, New York: Verso.

Battaglia, Debbora, David Valentine, and Valerie Olson. 2015. "Relational Space: An Earthly Installation." *Cultural Anthropology*, 30(2): 245–256.

Beaulieu, Anne. 2010. "Research Note: From Co-location to Co-presence: Shifts in the Use of Ethnography for the Study of Knowledge." *Social Studies of Science*, 40(3), 453–470.

Bennett, Tony. 1995. *The Birth of the Museum: History, Theory, Politics.* London: Routledge.

Bennett, Tony, Fiona Cameron, Nelia Dias, Ben Dibley, Rodney Harrison, Ira Jacknis, and Conal McCarthy. 2017. *Collecting, Ordering, Governing: Anthropology, Museums, and Liberal Government.* Durham, NC and London: Duke University Press.

Berry, Maya J., Claudia Chávez Argüelles, Shanya Cordis, Sarah Ihmoud, and Elizabeth Velásquez Estrada. 2017. "Toward a Fugitive Anthropology: Gender, Race, and Violence in the Field." *Cultural Anthropology* 32(4): 537–65.

Bichsel, Christine. 2020. "Introduction: Infrastructure on/off Earth." *Roadsides* 3: 1–6. https://doi.org/10.26034/roadsides-202000301.

Buchli, Victor. 2020. "Extraterrestrial Methods: Towards an Ethnography of the ISS." In *Lineages and Advancements in Material Culture Studies: Perspectives from UCL Anthropology*, edited by Timothy Carroll, Antonia Walford and Shireen Walton, 17–32. London: Routledge.

Candlin, Fiona. 2008. *Museums, Modernity and the Class Politics of Touching Objects in Touch in Museums, Policy and Practice in Object Handling*, edited by Helen J. Chatterjee. London: Routledge.

Chalk, Hannah-Lee. 2012. "Mobile Stones: The Uses and Meanings of Earth Science Teaching Specimens." *Material Culture Review* 74 (January): 149–160.

Chatterjee, Helen. 2008. *Introduction in Touch in Museums, Policy and Practice in Object Handling*, edited by Helen J. Chatterjee. London: Routledge.

Classen, Constance (Ed.). 2005. *The Book of Touch.* 1st ed. London: Routledge.

Critchley, Hugo. 2008. *Emotional Touch: A Neuroscientific Overview in Touch in Museums, Policy and Practice in Object Handling*, edited by Helen J. Chatterjee. London: Routledge.

de la Cadena, Marisol. 2014. "Runa: Human But Not Only." *HAU: Journal of Ethnographic Theory* 4(2): 253–259. https://doi.org/10.14318/hau4.2.013.

Daston, Lorraine. 2000. "Introduction: The Coming into being of Scientific Objects." In *Biographies of Scientific Objects*, edited by Lorraine Daston, 1–14. Chicago, IL: University of Chicago Press.

Dudley, S. 2011. "Encountering a Chinese Horse: Engaging with the Thingness of Things." In *Museum Objects: Experiencing the Properties of Things*, edited by S.H. Dudley, London: Taylor and Francis Group.

Drazin, Adam. 2020. "The Object Biography." In *Lineages and Advancements in Material Culture Studies*, edited by Timothy Carroll, Antonia Walford and Shireen Walton. London: Routledge. https://doi.org/10.4324/9781003085867-5.

Farrington, Oliver C. 1900. "The Worship and Folk-Lore of Meteorites." *The Journal of American Folklore* 13(50): 199–208. https://doi.org/10.2307/533884.

Fortey, Richard. 2000. "Introduction." In *Travels with the Fossil Hunters*, edited by Peter Whybrow, xiii–xvii. Cambridge: Cambridge University Press.

Gelsthorpe, David. 2021. "Decolonising Manchester Museum's Mineral Collection – A Call to Action." *Journal of Natural Science Collections* 9: 12–28.

Gorman, Alice C. 2005a. "The Archaeology of Orbital Space." In *5th NSSA Australian Space Science Conference 2005*, 338–57. RMIT University, Melbourne, 14–16 September.

Gorman, Alice C. 2005b. "The Cultural Landscape of Interplanetary Space." *Journal of Social Archaeology* 5(1): 85–107. https://doi.org/10.1177/14696 05305050148.

Gorman, Alice C., and Beth Laura O'Leary. 2013. "The Archaeology of Space Exploration." In *The Oxford Handbook of the Archaeology of the Contemporary World*, edited by Paul Graves-Brown, Rodney Harrison and Angela Piccini, 409–424. Oxford: Oxford University Press.

Gosden, Chris, and Yvonne Marshall. 1999. "The Cultural Biography of Objects." *World Archaeology* 31(2): 169–178.

Haraway, Donna. 1988. "Situated Knowledges: The Science Question in Feminism and the Privilege of Partial Perspective." *Feminist Studies* 14(3): 575–99.

Harjo, Laura. 2019. *Spiral to the Stars, Mvskoke Tools of Futurity*. Tuscon: University of Arizona Press.

Haskins, Caroline. 2018. "The Racist Language fo Space Exploration." *The Outline*. https://theoutline.com/post/5809/the-racist-language-of-space-expl oration

Hearth, S., and C. Robbins. 2022. "Mineral Displays as Embodiments of Geologic Thought and Colonial Invisibility." *Journal of Natural Science Collections* 10: 3–17.

Hicks, Dan. 2020. *The Brutish Museums. The Benin Bronzes, Colonial Violence and Cultural Restitution*. London: Pluto Press.

Hobart, Hiʻilei Julia. 2019. "At Home on the Mauna: Ecological Violence and Fantasies of Terra Nullius on Maunakea's Summit." *Native American and Indigenous Studies* 2(6): 30–50.

Knell, Simon J. 2000. *The Culture of English Geology, 1815–1851: A Science Revealed through its Collecting*. Aldershot: Ashgate.

Laing Ebbensgaard, Casper, Kerry Holden, and Kathryn Yusoff. 2022. "Planetary Portals in the Upside-Down." In *Working at the Intersection: Architecture After the Anthropocene*, edited by H. Harriss and N. House (eds) (Vol. 4). London: RIBA Publishing.

Lempert, William. 2021. "From Interstellar Imperialism to Celestial Wayfinding: Prime Directives and Colonial Time-Knots in SETI." *American Indian Culture and Research Journal* 45(1): 45–70.

Lester, Gustave, and Tamara Pico. 2021a. "The Colonial History of Geology in the United States, Part 1." *Insurrect!*. www.insurrecthistory.com/archives/2021/11/ 12/the-colonial-history-of-geology-in-the-united-states.

———. 2021b. "The Colonial History of Geology in the United States, Part 2." *Insurrect!*. www.insurrecthistory.com/archives/2021/11/17/the-colonial-history-of-geology-in-the-united-states-part-two.

Liboiron, Max. 2021. "Decolonizing Geoscience Requires More than Equity and Inclusion." *Nature Geoscience* 14: 876–877. https://doi.org/10.1038/s41561-021-00861-7.
McKinson, Kimberley. 2020. "Do Black Lives Matter in Outer Space?" *Sapiens*, 30 September. www.sapiens.org/culture/space-colonization-racism/.
Messeri, Lisa. 2016. *Placing Outer Space: An Earthly Ethnography of Other Worlds*. Durham, NC: Duke University Press.
Olson, Valerie. 2012. "Political Ecology in the Extreme: Asteroid Activism and the Making of an Environmental Solar System." *Anthropological Quarterly* 85(4): 1027–1044.
———. 2023. "Refielding in More-Than-Terran Spaces." In *The Routledge Handbook of Social Studies of Outer Space*, edited by Juan Francisco Salazar and Alice Gorman. London: Routledge.
Osbourne, Alana. forthcoming. *Decolonial Museology, Space Travel and the Mineral Cabinet*. Museum and Society.
Phillips, Rasheedah. 2018. "Placing Time, Timing Space: Dismantling the Master's Map and Clock." *The Funambulist*, Issue 18, 5 July.
———. 2019. "The Nowness of Black Chronopolitical Imaginaries in the Afro/Retrofuture." *The Funambulist*, Issue 24, 28 June.
———. 2021. "Counter Clockwise: Unmapping Black Temporalities from Greenwich Mean Timelines." *The Funambulist*, Issue 36, 21 June.
Pye, Elizabeth. 2008. *The Power of Touch: Handling Objects in Museums and Heritage Contexts*. Walnut Creek, CA: Left Coast Press.
Sage, Daniel. 2014. *How Outer Space Made America: Geography, Organization, and the Cosmic Sublime*. Farnham: Ashgate.
Smiles, Deondre. 2020. "The Settler Logics of (Outer) Space." *Society and Space*, 26 October. www.societyandspace.org/articles/the-settler-logics-of-outer-space.
Shorter, David Delgado and Kim TallBear. 2021 "An Introduction to Settler Science and the Ethics of Contact." *American Indian Culture and Research Journal* 45(1): 1–8.
Tilley, Christopher. 2004. *The Materiality of Stone: Explorations in Landscape Phenomenology*. Oxford: Berg Publishers.
Tsing, Anna. 2000. "Inside the Economy of Appearances." *Public Culture* 12(1): 115–144.
Valentine, David. 2016. "Atmosphere: Context, Detachment, and the View from Above Earth." *American Ethnologist* 43 (3): 511–524.
Yusoff, Kathryn. 2018. *A Billion Black Anthropocenes or None*. Minneapolis: University of Minnesota Press.
———. 2024. *Geologic Life, Inhuman Intimacies and the Geophysics of Race*. Duke University Press.
Yusoff, Kathryn, Kerry Holden, and Casper Laing Ebbensgaard. 2022. "Planetary Portals: 'Dreaming in Continents' from East London to the Cape in the Colonial Praxis of Emergence and Extraction." *Gropius Bau Journal*. www.berlinerfestspiele.de/en/gropiusbau/programm/journal/2021/kathryn-yusoff-planetary-portals.html

6 Of stars and wheat

Making sense of the cosmos in a regional museum of cosmonautics

Makar Tereshin and Denis Sivkov

After our late-night quest to see the monument dedicated to a Soviet cosmonaut[1] in an overgrown wheat field on the outskirts of Nikolaevsk, we returned to Kazmenko's family home to bring back the flashlight we had borrowed from them earlier. Since the late 1980s, the family has become known for heading the campaign for the establishment of a museum dedicated to Yuri Malyshev, the cosmonaut who was born in their hometown of Nikolaevsk. While we shared our impressions from the nocturnal adventure, Vladilen Korneevich Kazmenko invited us to explore their family archives. He retrieved a photograph of himself wearing Yuri Malyshev's spacesuit and standing at the entrance of the local museum "Earth–Kosmos."[2] The cosmonaut had generously allowed Vladilen Korneevich to try on the spacesuit and pose for a photograph. This was as close to outer space as Vladilen Korneevich could get – taking on the role of the cosmonaut and basking in the glory of the "space conqueror." In the photo, the figure of a newly minted cosmonaut stood on the doorstep of the local history museum in Nikolaevsk, in front of the "The Great Nomad" poster promoting a photo exhibition that was about to open in the museum (see Figure 6.1). In the late 1990s, when the photo was taken, the museum's workers were preparing a photo exhibition as a way to celebrate the agricultural cultivation and industrialisation of the Volga region steppes that paved the way not only for the steady progress of the Soviet Union but also, by extension, the emergence of the Soviet space industry.

Vladilen Korneevich began to recall his memories of Malyshev, telling us fondly about the last time he visited Nikolaevsk, his hometown. Suddenly, he lowered his voice conspiratorially. After a tour of the Soviet cities, Vladilen Korneevich said, the cosmonaut arrived in Nikolaevsk hungover, looking for another drink to help with his dizziness. Nina Alexandrovna, Vladilen Korneevich's wife and the main advocate for the museum's establishment, sharply interrupted her husband's story: "Enough! I don't like this! It's a betrayal!" She believed that this story "betrayed" the image of

DOI: 10.4324/9781003437956-6

Figure 6.1 Vladilen Korneevich in Malyshev's spacesuit, 1999. Courtesy of Museum Zemlya-Kosmos ('Earth-Kosmos'), Nikolaevsk.

a Soviet hero-cosmonaut, a man without bad habits, who set an example for the younger generation.

But for Vladilen Korneevich, Malyshev could be both at the same time; he was a hero and he was also a mate with whom he used to get drunk. Vladimir Korneevich told us that after the collapse of the USSR, Malyshev would often question the success and organisation of the Soviet space programme. This version of Malyshev was censored by Nina Aleksandrovna, who voiced her objections, saying that this was a private story relayed in confidence to Vladilen Korneevich: "Don't talk about it, don't! He told it to you in private, don't! ... Nobody is interested!" It was interesting to observe how the memory and narratives of the Soviet space programme and of one of its heroes, cosmonaut Malyshev, were breaking down in front of us into different competing versions.

Dozens of different cosmonautics[3] museums have emerged across the Soviet Union in the years following the first human spaceflight. Large museums of cosmonautics opened up in Moscow and Kaluga, forging the dominant code of language and visual representation of the Soviet space programme. Churches were turned into museums of space exploration and planetariums, both used for educating Soviet citizens about the social significance of space exploration. Small school exhibits and local history museums (*kraevedcheskie*[4]) in the homelands of prominent cosmonauts were also furnished with iconic images of the Soviet space programme in order to cultivate patriotic feelings towards the homeland amongst the locals and their sense of belonging to the greater Soviet project. Of these numerous locations, only the main state space museums and exhibits have received significant scholarly attention (Lewis 2005; Siddiqi 2005). In this chapter, we instead foreground Russian regional museums of cosmonautics as one of the key sites for the production of knowledge about space exploration. While, in part, reflecting the consolidated heroic narratives of space exploration and mirroring the iconic imagery of the main museums, these smaller museums provide an important comparative perspective on the Soviet and contemporary cultural practices of remembering the Space Age.

Foregrounding the Museum

In 2021, together with photographer Sergei Karpov, we decided to explore Russian regional museums and memorial complexes devoted to the Soviet cosmonauts and space exploration. Their regional status was of particular importance to us. Russia's regions are often contrasted with Moscow and Saint Petersburg, the country's political, economic and cultural centres, to emphasise their provincial or secondary status in a highly centralised country. Our main interest was in the alternative narratives about space exploration forged by regional institutions. We chose three museums: the Nikolaevsk Regional History Museum at the birthplace of cosmonaut Yuri Malyshev (Volgograd region), the memorial complex devoted to cosmonaut Andriyan Nikolayev in Shorshely (Chuvash Republic), and a museum of cosmonautics at a school for children with hearing disabilities in Pytalovo (Pskov region).

The idea of exploring Russia's regional museums emerged spontaneously. We were interested in working together on a collaborative project exploring the relatively unknown space-themed museums and memorials in Russian provinces. At first, the three of us simply exchanged photos from the Internet, compiling a list of places we would like to visit. We eventually narrowed it down to these three museums, which we felt stood out when compared to the conventional space museum. The deciding factor was our sincere surprise. Even though two of the three of us had lived in Volgograd for significant periods of our lives, neither one had ever heard that one of the Soviet cosmonauts came from the Volgograd region. The museum in Shorshely struck our imagination with the presence of an Orthodox chapel as part of the museum memorial complex, and in Pytalovo there was a museum created by and for deaf or hard-of-hearing children.

Despite the didactic nature of their expositions and the apparent resonance with the aesthetics of the cosmonautics museums in Moscow and

Kaluga, the creation of a local museum collection is a creative opportunity for museum staff, who most often lack professional museological experience. We discovered that they incorporated their own ideas and perceptions of the Soviet space heritage into the work of the museum. Their expositions were developed thanks to the private efforts and investments, relying on local initiatives of history enthusiasts (*kraevedy*). The support and networks of former cosmonauts, who were born and grew up in these places, also contributed to this process, allowing museum workers to travel to the central museums themselves and obtain the desired items. Navigating the hallways of Moscow's museum of cosmonautics, looking for inspiration, or petitioning space industry officials for support, museum workers were adopting conventional imaginaries and narratives of the Soviet space culture (Kohonen 2017; Gerovitch 2015; Maurer et al. 2011; Andrews and Siddiqi 2011; Collins and Millard 2005). But they were also able to bring the space culture back home, where the history of space exploration was embedded in the history of their own region, foregrounding the importance of its role in the workings of the national and global space industries.

In anthropological scholarship, museums have been widely theorised as contact zones. This concept was coined by Mary Louise Pratt (1991) and then later elaborated by James Clifford in relation to museums. Clifford (1997, 213) proposed to "view [...] all culture-collecting strategies as responses to particular histories of dominance, hierarchy, resistance, and mobilisation." In this light, museums are seen as politicised spaces, ideologically charged in how they engage and represent particular communities and their past, present, and future. Other scholars have highlighted how practices of collection, classification, and display both produce and represent our knowledge about the world (Geismar 2018). They are also embedded in the governance of populations promoting dominant values and ideologies (Bennett et al. 2017, 1995). Taken to its limit, this view regards museums as weapons that legitimise, extend, and naturalise, new extremes of colonial violence, silencing histories of loss and death (Hicks 2020, 15). Struggles for inclusion, sharing of authority, or the restitution of cultural property, are at the heart of the contemporary debates about museum collections (von Zinnenburg Carroll 2022; Giblin, Ramos, and Grout 2019; Boast 2011; Clifford 2004; Cuno 2008). However, museums can also be sites of critique and contestation. As Boast suggests (2011, 65), Indigenous museums and cultural centres can create their own centres of collecting, performance, and presentation, appropriating the technology of museums to their own ends. Likewise, Murawska-Muthesius and Piotrowski (2015) argue that museum collections and their cultural authority can be used to take a stance on the most pressing issues and articulate contemporary injustices.

In this chapter, we take a close look at one of the three museums we explored in our research – the museum of regional history "Earth–Kosmos" in the town of Nikolaevsk. Focusing on the key imaginaries promoted by the museum, this chapter explores how this museum represents the Soviet space legacy for contemporary audiences in Russia's provinces. In capturing the ways in which the museum workers collapse scales and present the cosmic for their visitors, we specifically uncover the methodological toolkit of *cosmovedenie* – space studies as conceived by workers of regional history museums in Russia. At the heart of *cosmovedenie* is the Earth, people who inhabit it, the feeling of belonging, and a sense of unity among the people sharing a homeland in which they were born.

Introducing the museum collection, community, and narratives, we demonstrate how space can be ethnographically situated and engaged with in and through the study of a museum. Through an ethnography of the museum exhibits and space memorabilia, which are devoted to the live of Yuri Malyshev, a fellow cosmonaut from Nikolaevsk, we demonstrate how this museum entangles the global imagery of the Soviet space with intimate local histories and personal perceptions of the country's past and future.

Methodology

We began our project in 2020, but had to wait to conduct fieldwork until museums re-opened after the first wave of Covid-19 lockdowns. Constrained by our funding and timeframes for research deliverables, we were limited to three brief but intense expeditions to each of the locations in between Covid lockdowns. With such a short period of time, we could not afford to embark on long-term fieldwork, so we built our ethnography around rapid (Vad Karsten 2019) and multimodal (Collins and Durington 2017; Pink 2011; Dicks et al. 2006) methods. These methodological approaches help us to question the conventional research pace and timeframes in a dramatically changed fieldwork environment. Rather than waiting for an "ethnographic moment" (Strathern 1999), we treated everything that happens in the field as potential data. In this sense, the methodological and epistemological boundaries between a visit, field report, photography, video, observation, interview, and archival work were blurred. The three of us tried to seize all opportunities, to capture all words and speeches, to be in one place together or in several places at once, to experience the same event in different ways, to complement questions and answers, to learn from each other and share our different skills and ways of seeing.

The reality of the museum that was available to us during fieldwork can be thought of as an ongoing museum tour, but was not limited by it. The museum workers treated us – the researchers who came from two capital cities, Moscow and St Petersburg – as conduits for their story, who would tell the people "back there, in Moscow" about their beloved museum, the people who stood behind its work, and the place they were intimately connected to. In this sense, during the excursions (both formal and informal) that we were part of, the identity of a community was reassembled and manifested in, through, and around, the museum exhibits. That said, the narrative of the tour was not unified or consolidated; it was embodied in different voices and disintegrated into separate narratives that were sometimes censored and corrected by other museum workers.

From *kraevedenie* to *cosmovedenie*

The first space exhibition in the Soviet Union, and most likely in the world, was opened in Moscow in 1927. It was called the *World Exhibition of Models of Interplanetary Vehicles and Mechanisms* and was initiated from "below." The exhibition was organised without state support by enthusiasts from the Association of Inventors and included the literature on space flights, models, and drawings of flying machines. It attempted to tell people about something as of yet unknown but strangely appealing – something no one had ever seen – and focused on the practicalities of hypothetical spaceflight. In addition to the Soviet scientists Konstantin Tsiolkovsky and Friedrich Zander, contributors to the exhibition included international pioneers of space exploration, namely Robert Goddard, Hermann Oberth, Max Vallier, Robert Esnault-Peltri, and others (Siddiqi 2010, 92–97). Space enthusiasts who dreamt about transcending Earth used the exhibition as a way to educate lay audiences on space technologies and promote space exploration.

Three decades later, Soviet museums and pavilion exhibits established after Garagin's flight followed a different logic. They represented the, by then active and successful, Soviet space programme through replicas of spacecrafts and satellites, while original artefacts were exhibited behind the closed doors of the Soviet space industry centres (Gerovitch 2015, 17; Lewis 2005; Siddiqi 2005). Instead of revealing the technical, most often classified, information from the history and development of Soviet space technologies, these exhibits called for an affective response from the public by displaying a future-oriented idealised version of reality (Lewis 2005, 152). As part of the Soviet space myth, vividly described by Slava Gerovitch (2015, xv), they "gave tangible representations to the ideological concepts of socialism and nationalism, and cemented the identity of a nation."

Although not completely monolithic, the Soviet space myth was nevertheless exclusive, with only certain people allowed to contribute to its construction. As Gerovitch (2015, 22–26) notes, cosmonauts, space engineers, and military officials all contributed their own ideas to the space mythology and participated in reshaping or eroding the official narratives to suit their own agendas. This fragmentation was further exacerbated during perestroika and after the collapse of the Soviet Union. Corporate museums opened up for ordinary spectators, previously unseen space artefacts were put on public display, key figures of the Soviet space industry started to publish their memoirs, while the archives were opened up for researchers (Siddiqi 2005; Gerovitch 2015, 157). As Siddiqi puts it, "the single narrative of Soviet space history – teleological and whiggish – fractured into multiple and parallel narratives full of doubt (for the claimed successes of the programme), drama (for the episodes we never knew about) and debate (over contesting narratives of history)" (2005, 99). Regional cosmonautics museums and *kraevedcheskie* were also active in reshaping these fragmented stories.

Regional museums or museums of local history, known as *kraevedcheskie*, were one of numerous institutions that took part in the formation of the post-Soviet space culture in Russia. *Kraevedcheskie* museums emerged from amateur research activities in the nineteenth-century Russian Empire. The activities of *kraevedcheskie* museums were part of the wider efforts to promote public education working with local material and oriented towards the social and economic development of a particular territory (Gavrilova 2022; Smirnov 2018). After the revolution, these local initiatives were incorporated into a large state-sponsored project of accumulating data on regional geography, ethnography, and history, for the needs of the socialist planned economy, governance, and knowledge production. The institutional framework of *kraevedenie* spawned numerous small regional research centres, often established within local museums spread all over the Soviet Union. With the collapse of the Soviet Union, this centrally planned museum infrastructure was eroded but did not cease to exist.

Given the need to reflect technological advancement of the Soviet Union and stimulate popular support for the Soviet state, some of these museums positioned themselves in relation to cosmonautics and began introducing small exhibits on the role of the region in the Soviet space programme. However, despite the ubiquity of the Soviet space culture, not every museum had a chance to furnish their expositions with exquisite space-flown artefacts, or even with their replicas. Most of the objects flown to space and cosmonauts' memorabilia are kept by the Russian Space Agency and redistributed across the more privileged Russian museums of cosmonautics rather than shared with their regional counterparts across Russia.

When compared to other space-themed museums, *kraevedcheskie* museums stand out in their approach to curating and showcasing space.

Although *kraevedeniye* could be characterised as originating from a governmental modernist project and Soviet institutional framework, functioning according to strict guidelines (Gavrilova 2022), the work of such museums is also determined by local conceptions of regional culture, history, and science, as well as by the regional repercussions of the national political agenda and ideologies (Melnikova 2015; Donovan 2019). Therefore, the workers of space-themed *kraevedcheskie* museums utilise the Soviet space heritage, establishing fragile connections between their home region and outer space. These links not only expand the local place. The mix of space and regional collections in the museum augment *kraevedinie*'s horizontal dimension of a small town with the verticality of the *cosmovedenie* that connects it with other places in the universe.

Our very own cosmonaut

Yuri Vasilyevich Malyshev is the first and only offspring of the Volgograd region who has flown to space. His first space flight took place on 5 June 1980. According to Soviet custom, the cosmonaut was awarded the title of hero of the Soviet Union, the nation's highest honour. Four years later, Malyshev made a second flight and became a two-time hero of the Soviet Union. In accordance with the Soviet protocol, when one received their second award, a bronze bust of this person had to be installed in the hero's hometown. In the case of Nikolaevsk, where Malyshev was born, this was not a straightforward task. Just as many other towns and rural areas across the USSR, old Nikolaevsk was flooded during the construction of the Volga Hydroelectric Station in the 1960s. The lower part of the present-day town, former *sloboda* (settlement) Nikolaevskaya, was submerged in the reservoir, while its inhabitants were resettled further away from the river to an emerging town which was named after the old town. When the question of where to place the bust of Malyshev came up, local *kraevedy* and cultural workers, supported by the district authorities, decided to seize the moment and lobby for the creation of a local history museum that would tell the story of their fellow countryman cosmonaut and of the relocated town where he grew up.

Late Soviet democratisation and political reforms provided a chance to further interests of the local *kraevedy* who wanted to address the rupture of the social fabric wrought by the forced resettlement. Most of Nikolaevsk *kraevedy* and cultural workers, who were behind the establishment of the museum, came from the Soviet state cultural and educational sector, where they worked as librarians, schoolteachers, or as activists and members of the Communist Party youth organisation. They formed the most active part of the museum advocates in Nikolaevsk. In 1986, the local enthusiasts managed to obtain permission to convert the building of the

Figure 6.2 Space Hall of the "Earth–Kosmos" museum, 2021. Courtesy of Sergey Karpov.

local department store, still under construction at the time, into a museum. In an attempt to incorporate local history into the centralised patriotic discourse, they secured state support cementing the museum's subordination to the regional Ministry of Culture. In 1990, the *kraevedchesky* museum "Earth – Kosmos" welcomed its first visitors (Figure 6.2).

Today, museum employees are paid from the municipal budget. This is considered a good job given the overall economic instability in Russia's small towns. It guarantees a small but steady salary which can be coupled with informal income from private land plots, seasonal jobs, and social benefits. Furthermore, one can use their position to help relatives and acquaintances to get a job at the museum. But for many, the work of the museum also holds ideological significance. In the 1990s, against the backdrop of collapsing state ideology, local cultural workers felt a strong moral imperative to provide pedagogical guidance through the examples set by the hero-cosmonauts, tapping into continuous popular fascination with outer space. This impulse has strengthened and solidified over the past decades, given renewed attempts to establish official state ideology, specifically when it comes to the Soviet legacy and the Great Patriotic War.

Mobilising their personal relations and connections to the cosmonaut, *kraevedy* established an alliance built on the sense of belonging.

In their eyes, Malyshev emerged from their shared homeland with its inherent richness forging his destiny. Despite leaving his home behind, Malyshev remains forever tied to the native region and thus his return is inevitable. Cosmonauts return to their homeland to socialise with their fellow countrymen and to recover their strength. As Malyshev's classmate Antonina Vasilyevna said about the cosmonaut: "Gagarin was the first, but Malyshev – he is our own, he is local ... He is ours. I know him, and there he is." In such affirmations of the local identity, particular emphasis is placed on the local. In this context, the word "native" alludes to the small place one comes from, rather than to the country of origin. Our interlocutors often spoke of their fellow cosmonaut Malyshev as "*zemlya*," describing him as someone who shares the same soil (the same earth) with them. As such, *zemlyak* has the same root as *Zemlya*, meaning Earth. But *zemlyak* is not the same as Earthling, someone from the Earth, it is specifically someone who belongs to the same place as you. In that sense, Earthlings (*zemlyane*) become *zemlyaks*.

Exhibiting space

Now, let us take a closer look at the exhibition halls. The museum is divided into three halls, each offering a unique perspective on Nikolaevsk's history. The first, known as the Space Hall, delves into the narrative of the Soviet space programme and Malyshev's spaceflight. Moving on, the second hall provides a detailed exploration of Old Nikolaevka, offering insights into the daily life of the nineteenth-century town. The final hall, a recent addition, sheds light on the role of Nikolaevka during the Great Patriotic War (1941–45), detailing its contributions to the battles for Stalingrad and the regional partisan movement.

As we enter the museum, the figure of Icarus storming the sky opens up the diorama running throughout the exhibition hall. Icarus is followed by representations of early astronomical theories, navigation tools, and portraits of astronomers – Galileo, Copernicus, and Lomonosov. The next section depicts the "space trinity" – Tsiolkovsky, Korolev, and Gagarin – followed by Tereshkova, Malyshev, and Rakesh Sharma, Malyshev's fellow cosmonaut on the Intercosmos mission. Together with our tour guide, the museum leads us through the Soviet space myth, through the history of the space programme to Nikolaevsk's own cosmonaut. The rest of the diorama represents the link between space exploration and the national economy and technological development, traditional for representations of the Soviet space programme. The focal point of the composition depicts a state-of-the-art combine harvester, driving through meticulously rendered ears of wheat. Above, futuristic spaceships soar through the cosmic expanse, all set against the starry skies and distant planets.

Here, it is easy to notice the connection between technical achievements in space and technical solutions on Earth. In the fragment of the diorama, space exists not for the utopian dream of leaving Earth and starting from scratch, but for the earthly needs of the Soviet citizens. In this sense, graphically, combine harvester and spaceship, wheat and stars rhyme with each other well. Mastering space allows for observing the Earth's climate and weather from orbit, which in turn allows for better cultivation of the land and higher yields. Altogether, it delivers a clear message to the visitors – space exploration manifests and mirrors the technological advancement of the socialist system (see Figure 6.3).

The "economic" block has an additional meaning. Curiously, immediately after the concluding agrarian scenes of the Space Hall, the visitor walks into the local history hall, three or four times smaller in size, where the history of the region is represented. A model of an old church demolished during the construction of the Volga power plant is reproduced in the centre. It is surrounded by the scenes of urban life at the turn of the nineteenth century. The rest of the hall showcases the gifts of the Earth, such as wheat, salt from Lake El'ton, and watermelons. The transition from the history of space exploration to the local seems almost seamless. This feeling is retroactively reinforced by the photos of Yuri Malyshev holding wheat in a collective farm's field pictured in the

Figure 6.3 Sketches for the Space Hall diorama, 2021. Photograph by Denis Sivkov.

Space Hall. After his flight, the cosmonaut was accepted into the team of mechanics and machine operators in one of the district farms. The hay field assigned to Malyshev was also named after him, and a stele depicting the cosmonaut's head, the stars, and, of course, more wheat, was installed on this (his) field. The stele and the field itself have become overgrown with acacia, but the place is still called "Malyshev's field" to this day (see Figure 6.4).

Thus, through this visual language, the Nikolaevsk museum begins to tie together Nikolaevsk, the place where the cosmonaut grew up and nourished himself with the gifts of his native land, and the distant unexplored space that Malyshev helped to make closer and more familiar. Such representations of the space frontier in small space museums highlight a

Figure 6.4 Yuri Malyshev inspecting a collective farm during harvesting season, 1980s. Courtesy of Museum Zemlya-Kosmos ("Earth–Kosmos"), Nikolaevsk.

linguistic peculiarity to which we would like to draw your attention. These small museums are often located in regions and places that in Russian are referred to as *krai*. In Russian, the word *krai* has a multifaceted meaning. Its literal translation is edge, border, or frontier. Prosaically, it is used to refer to the country's borderlands, its physical and symbolic edges; but in a more abstract sense, it also invokes a deep emotional and cultural connection to a particular region or place within the country. Here, the cosmonauts are spoken of as fellow countrymen and natives of this region. People say that their native land, their *krai*, nurtured them, and made them what they became.

Krai is not exclusively the border of an empire, not just the horizon of the infinite Universe, but an intimate, animated, limited place where, through the space museum, perception is expanded and the entire Universe becomes accessible. *Krai* in the *kraevedcheskiy* museum is represented by the people, communities, and things. *Krai* is inhabited by fellow countrymen who simultaneously live in their town or village, on the planet, and in the infinite Universe. Small space museums do not reduce the infinite space to an earthly limited place (Messeri 2016). They do not simply connect places of different scales on Earth and in space (Olson 2018). Rather, space museums amplify earthly places – in the museum, these places are actualised and made visible precisely against the backdrop of the immeasurable final frontier. Space entrapped within a museum spotlights a small town, neighbourhood, or village, turning them into a visible and special place, unlike other places.

But it is not only localisation that is important. The juxtaposition of scales, in which the infinite and immense Universe connects with the small and commensurate *krai*, and the distant becomes close, is also expressed in the involvement of Nikolaevsk residents in space exploration. In Nikolaevsk, we observe the work of "continuous arrangements of scalar relations" described by Valerie Olson (2018, 31). She suggests that the notions of systematicity and imaginations of environmental relations used by engineers and others in the American space programme allow different places and scales, or places of different scales, to be connected and juxtaposed. Using the example of Nikolaevsk, we see that this kind of continuous arrangement of scalar relations is not limited to space specialists.

This rhetoric of connecting the local earthly places and the immense space could be found throughout the museum exposition and among visitor reviews. Visitor logbooks provide a valuable entry point into understanding the public's interaction with the museum, articulating not only the formal representations within exhibits but also the informal, lived experiences of those who engage with the museum environment. In the following response by a museum visitor, presented in its entirety, we can clearly see how the planetary and the local are combined in one statement, while the vast cosmos is located in a specific place:

> We are all inhabitants of the cosmos living on our marvellous planet Earth. And how great it is that the Nikolaevsk Museum is called simply and majestically "Earth – Kosmos." Thank you Natalia Anatolievna, the director of this museum, for your careful attitude to the history of Nikolaevsk, and most importantly to its people who glorified this beautiful land of the Volga region.
>
> (A. Kuprin, May 2013)

Despite the formal and ritualistic nature of such responses, they contain curious spatial and temporal markers. Moreover, these markers are connected with the various scales of space exploration: the activity of all of mankind (global scale), the result of the space race of states (national scale), or the expression of actions of individuals and small groups – communities (local scale). Analysing the book of reviews, we saw that visitors coming to Nikolaevsk from large cities, such as Moscow, Samara, Volgograd, when thanking the museum, tend to place it within the framework of the national spatial categories. In their reviews, space exploration is carried out and represented by the Russian people in the country of Russia. At the same time, residents of Nikolaevsk itself and, interestingly, of similar small settlements note something different; in their responses, the emphasis is not on the national, but on the local – on the region and their native land. In these and other reviews, the space is localised not in the country, but in the Volga steppes, in one's home region.

These responses connect small towns and villages with distant, unknown, and large expanses of the Universe, making it closer for museum visitors, not only scientists, engineers, and cosmonauts. On one hand, space acquires a new density of meaning: not merely earthlings, but cherished fellow countrymen appear in it, and the universe turns into a *krai*. At the same time, through proximity and acquaintance with fellow countrymen-cosmonauts, their *zemlyaki* find themselves involved in something more than their immediate lives.

Conclusion

In the context of the ethnographic study of outer space, looking at museums offers a chance to consider cosmonauts through a localised perspective. The study and exploration of space is not only carried out by the "envoys of mankind" (UNOOSA, 1966), scientists, and engineers; it also involves a regional museum's visitors in places far from the centres of scientific and cultural production known to all. Therefore, it is worth looking down to

Earth beyond the imaginaries and views-from-nowhere, following more closely the processes and routes that people follow in order to imagine and make sense of the cosmos otherwise.

Exploring exhibitions and object collections inside a museum, its architecture, layout, decorative murals, and visitor logbooks, are multiple points of entry into the social life of the museum. These shed light on how contemporary space cultures are created and reproduced through different media and modes of engagement with museums' material environment, and how material structures of the museum produce and communicate meanings to visitors and museum workers as they interact with them.

Acknowledgements

We are very grateful to Sergey Karpov. This project would not have been possible without our collaboration. The work of the authors was supported by the Centre for Experimental Museology Grant Programme "Museum in a liminal state," V-A-C Foundation. Makar Tereshin's work is supported by the UK Arts and Humanities Research Council (grant number AH/R012679/1).

Notes

1 A term originating from the Soviet space programme, referring to an astronaut. Cosmonauts are individuals trained to operate spacecraft and conduct missions beyond Earth's atmosphere.
2 In Russian, *kosmos* refers both to outer space and cosmos, which Earth is a part of.
3 In Russian, the word "cosmonautics" refers to the theory and practice of space travel, encompassing the science, technology, and activities involved in sending humans and spacecraft into outer space.
4 Throughout the text we are using related terms. They are all related but one is an adjective for the museum (*kraevedchesky*), one is the practice (*kraevedenie*), and one is the person or people carrying out the practice (*kraeved* or *kraevedy*).

References

Andrews, James T, and Asif A Siddiqi. 2011. *Into the Cosmos: Space Exploration and Soviet Culture*. Pittsburgh, PA: University of Pittsburgh Press.

Bennett, Tony. 1995. *The Birth of the Museum. History, Theory, Politics*. London: Routledge.

Bennett, Tony, Cameron, Fiona, Dias, Nelia, Dibley, Ben, Harrison, Rodney, Jacknis, Ira, and Conal McCarthy. 2017. *Collecting, Ordering, Governing: Anthropology, Museums, and Liberal Government*. Durham, NC and London: Duke University Press.

Boast, Robin. 2011. "Neocolonial Collaboration: Museum as Contact Zone Revisited." *Museum Anthropology* 34, no. 1: 56–70.
Clifford, James. 1997. "Museums as Contact Zones." In *Routes: Travel and Translation in the Late Twentieth Century* edited by James Clifford, 188–219. Cambridge, MA: Harvard University Press.
———. 2004. "Looking Several Ways: Anthropology and Native Heritage in Alaska." *Current Anthropology* 45, no. 1: 5–30
Collins, Samuel Gerald, and Matthew Slover Durington. 2017. "Multimodality: An Invitation." *American Anthropologist* 119, no. 1: 142–146.
Collins, Martin and Douglas Millard. 2005. *Showcasing Space*. London: Science Museum.
Cuno, James. 2008. *Who Owns Antiquity? Museums and the Battle Over Our Ancient Heritage*. Princeton, NJ: Princeton University Press.
Dicks, Bella, Bambo Soyinka, and Amanda Coffey. 2006. "Multimodal ethnography." *Qualitative Research* 6, no. 1: 77–96.
Donovan, Victoria. 2019. *Chronicles in Stone: Preservation, Patriotism, and Identity in Northwest Russia*. Ithaca, NY: Cornell University Press.
Gavrilova, Sofia. 2022. *Russia's Regional Museums: Representing and Misrepresenting Knowledge about Nature, History and Society*. London and New York: Routledge, Taylor and Francis Group.
Geismar, Haidy. 2018. *Museum Object Lessons for the Digital Age*. London: UCL Press.
Gerovitch, Slava. 2015. *Soviet Space Mythologies: Public Images, Private Memories, and the Making of a Cultural Identity*. Pittsburgh, PA: University of Pittsburgh Press.
Giblin, John, Ramos, Imma, and Nikki Grout. 2019. "Dismantling the Master's House: Thoughts on Representing Empire and Decolonising Museums and Public Spaces in Practice. An Introduction." *Third Text* 33, no. 4–5: 471–486.
Hicks, Dan. 2020. *The Brutish Museums: The Benin Bronzes, Colonial Violence and Cultural Restitution*. London: Pluto Press.
Kohonen, Iina. 2017. *Picturing the Cosmos: A Visual History of Early Soviet Space Endeavor*. Bristol: Intellect Books.
Lewis, Cathleen. 2005. "The Birth of the Soviet Space Museums: Creating the Earthbound Experience of Space Flight During the Golden Years of the Soviet Space Programme, 1957–68." In *Showcasing Space* edited by Martin Collins and Douglas Millard, 143–144. London: Science Museum.
Maurer, Eva, Richers, Julia, Rüthers, Monica and Carmen Scheide, (eds.) 2011. *Soviet Space Culture: Cosmic Enthusiasm in Socialist Societies*. London: Palgrave.
Melnikova, Ekaterina. 2015. "Lokalnoe Proshloe na Granitse Natsionalnogo: Kraevedcheskie Muzei Severnogo Priladozhya v Postsovetskoe Vremya." *Antropologicheskiy Forum* 25: 9–41.
Messeri, Lisa. 2016. *Placing Outer Place: An Earthly Ethnography of Other Worlds*. Durham, NC: Duke University Press.
Murawska-Muthesius, Katarzyna, and Piotr Piotrowski, (eds.) 2015. *From Museum Critique to the Critical Museum*. Surrey: Ashgate Publishing.

Olson, Valerie. 2018. *Into the Extreme: U.S. Environmental Systems and Politics Beyond Earth*. Minneapolis: University of Minnesota Press.

Pink, Sarah. 2011. “Multimodality, Multisensorality and Ethnographic Knowing: Social Semiotics and the Phenomenology of Perception.” *Qualitative Research* 11, no. 3: 261–276.

Pratt, Mary. 1991. “The Arts of the Contact Zone.” *Profession* 91: 33–40.

Siddiqi, Asif. 2005. “Privatising Memory: The Soviet Space Programme Through Museums and Memoirs.” In *Showcasing Space* edited by Martin Collins and Douglas Millard, 98–115. London: Science Museum.

———. 2010. *The Red Rocket's Glare: Spaceflight and the Soviet Imagination, 1857–1957*. Cambridge: Cambridge University Press.

Smirnov, Nikolai. 2018. “Mnozhestvo i Odin krayevedcheskiy muzei: ot mnogoobraziya alternativnykh proektov do lokalizatsii edinogo narrativa sovremennosti.” In *Almanakh TSEM: Nomer 1* edited by Ekaterina Shestova, 301–342. Moscow: V–A–C Press.

Strathern, Marilyn. 1999. *Property, Substance, and Effect: Anthropological Essays on Persons and Things*. London: Athlone Press.

UNOOSA (United Nations Office for Outer Space Affairs). 1966. “Treaty on Principles Governing the Activities of States in the Exploration and Use of Outer Space, including the Moon and Other Celestial Bodies.” 18 December. www.unoosa.org/oosa/en/ourwork/spacelaw/treaties/introouterspacetreaty.html.

Vad Karsten, Mette Marie. 2019. “Short-term Anthropology: Thoughts from a Fieldwork Among Plumbers, Digitalisation, Cultural Assumptions and Marketing Strategies.” *Journal of Business Anthropology* 4: 108–125.

von Zinnenburg Carroll, Khadija. 2022. *The Contested Crown: Repatriation Politics Between Europe and Mexico*. Chicago, IL and London: University of Chicago Press.

7 From Mexico to the Moon

(Outer)spatialising ethnography

Anne W. Johnson

As an organising thematic principle, this chapter centres on Mexico's relation to the Moon. The Moon featured prominently in pre-Hispanic myths and legends and continues to inspire contemporary artworks. But Mexico has also become a participant in the global outer space industry, through its role in NASA's Artemis project, as well as through emerging technoscientific activities and the insertion of Mexican actors into the NewSpace economy. In addition to providing methodological insights, the Moon missions discussed in this chapter lead to an examination of Mexico's relationship with the Moon from the past into the future.

Mexico and the Moon

Rosa Inés: What was your first memory of space?
Italia: Umm, the Moon.
Rosa Inés: The Moon? And how did it look?
Italia: Umm, like a big round ball.
Mariel: And what do you think it's like out there in space?
Italia: Well, I think it's really big.
Rosa Inés: If you could go to the Moon or space, right now, would you go?
Italia: Yes, but only if I could go with my mom.[1]

Rabbit in the Moon

Rosa Inés: What was your first memory of space?
Joshua: I think it must have been the song about the Moon, the one about the rabbit. Every full moon my mother sang it to me. That must have been it.[2]

Before day and night existed, the gods gathered in Teotihuacan to decide how to illuminate the world. Two gods, the wealthy Tecuciztecatl and the poor and sickly Nanauatzin, offered to sacrifice themselves to become

DOI: 10.4324/9781003437956-7

the Sun and the Moon. When it came time to throw themselves onto the pyre, Tecuciztecatl hesitated: "He exerted himself to the full, that he might cast and give himself to the flames. And he could in no way dare to do it. When again the heat reached him, he could only turn and leap back. He could not bear it." But Nanauatzin "had no fear, he did not stop short; he did not falter in fright; he did not turn back. All at once he quickly threw and cast himself into the fire ... Thereupon he burned; his body cracked and sizzled." Tecuciztecatl was ashamed, so he threw himself into the fire and burned as well. At first, both burned so brightly the other gods could not bear to look up, so they decided that the celestial bodies must be differentiated: "Then one of the gods came out running. With a rabbit he came to wound in the face this Tecuciztecatl; with it he darkened his face; he killed its brilliance. Thus doth it appear today" (Sahagún 1950, 4–7).[3]

According to noted scholar of Mesoamerica Alfredo López Austin, the rabbit in the Moon was associated with *pulque*,[4] the southern cardinal direction, cold things, intoxication, fermentation, menstruation, and pregnancy (López-Austin 1996, 5). The lunar calendar was also in widespread use among pre-Hispanic cultures; the ancient Maya, for example, calculated time with reference to the movements of both the Sun (years of 365 days divided into 18 months of 20 days each) and the Moon (lunar months, divided into alternating 29- and 30-day synodic cycles). The lunar count was beneficial for orienting agricultural activities and continues to be in use in some contemporary Mayan communities (Iwaniszewski 2012, 40).

In 2018, the Mexican Space Agency (AEM) sponsored a space art contest for young people and adults, marking the announcement of NASA's Artemis programme with the theme "Mexico to the Moon." The winning entry, "Dreams are built on roots and pursued with determination and preparation," depicts a young Indigenous girl climbing a ladder that emerges from a pyramid up towards the Moon (see Figure 7.1). Inside the pyramid, the artist has reproduced the famous imagery from an engraving in the Mayan ruins of Chichén Itzá of Pakal the Great, better known as "The Traveller" or "The Astronaut" for what appear to be vines that extend from the central figure's body, interpreted by some as the tubes or cables of a space ship. The Moon, a smiling inclined crescent, is accompanied by a rabbit and the words "In the navel of the Moon, we have always been here."[5]

Imaginary Voyages

Mariel: Is there a particular feeling or emotion that you have when you remember the stars or see the planets?

Figure 7.1 "Los sueños se construyen sobre las raíces" by Juan Carlos Cuevas Méndez. Courtesy of the Mexican Space Agency.

Rey Humberto: The thing is, we're just a tiny part of the Universe. We're nothing. The universe is so big, and there are so many things to discover. And sometimes you just think, like astronauts, or everyone, really, "Well, where do we come from? What are we doing here?" And, wow, no! Because you look and you say to yourself, "Well, when will I go to the Moon?" We'll never have the chance to see all of those wonders, right? Or, maybe, in the future. But it isn't for everyone. Just the ones that have the means are the

> ones that will be able to travel because those trips are really expensive.[6]

The Franciscan friar Manuel Antonio de Rivas arrived in Yucatán, Mexico in 1742 from Spain. He is best known for a proto-science-fiction story with the cumbersome title *Syzygies and Quadratures of the Moon Arranged to Merida of Yucatan's Meridian by an Anctitone or Moon Inhabitant, and Addressed to Bachelor Ambrosio De Echeverría, Deacon of Funeral Kyries at the Parish of Jesus of Said City, and at Present Professor of Logarithms in the Village of Mama on the Yucatan Peninsula, in the Year of Our Lord of 1775*. In the narrative, in which the author imagines an exchange of scientific information between terrestrial and lunar scientists, a French astronomer called Dutalon educates the Moon dwellers on the theories of Descartes and Newton. In turn, they express their doubts about biblical messianic teleology, and inform Dutalon that Hell is located in the Sun. The lunar scientists calculate the position of Merida, Yucatán's capital city, and discover that as the Earth spins rapidly at that latitude, its inhabitants must suffer from "permanent vertigos or fainting spells that impede the functioning of rational souls [...] They say that yes, we are people, yes, but what kind of people? People without speech, without shame, without brains, frauds, inconstant, lunatics. Look who's talking!" (Rivas et al. 2009 [1775], 60). Rivas presents the Moon, in contrast to the Earth, as a utopia, without the injustices and inequalities that characterise terrestrial society. In 1777, Rivas was put on trial for heresy, but the charges were dropped after the Inquisition determined that the tale should be considered a fable.

Between 1965 and 1970, astronauts participating in the Apollo programme were sent by NASA to the Biosphere Reserve of El Pinacate in north-western Mexico. Sacred to the Tohono O'odham, the reserve's lunar-like landscape is highly valued by biologists for its diversity of plant and animal life, and by geologists for its immense dunes and lava fields. Crew members of the Apollo 14 mission "were there for five whole days from April 14 to 18 1970. During their time, they laboured from 8 am until 6 or 7 pm simulating the different tasks and hikes they would undertake during their working days on the Moon" (Arreola Santander 2017). Perhaps for this reason, Mexico was one of the 135 countries to receive the gift of a Moon rock, on permanent display at Universum, the science museum at the Universidad Nacional Autónoma de México (UNAM).

In 2016, artist and photographer Juan José Díaz Infante transmitted fragments of *Don Quixote*, written by Rivas' contemporary Miguel de Cervantes, into space by bouncing them off the Moon using a shortwave radio and a Yagi antenna. For Díaz Infante, Quixote, who isn't afraid to dream the impossible, sends a "message that humanity should communicate to other civilizations in outer space," transcending Voyager, which "forgot to include goodness and messages of peace on its golden disk" (Aquino 2016).

That same year, Nahum created the project "Voyage: A Session for Remembering," in which the musician/artist/magician planted the memory of a trip to the Moon in participants' minds using hypnosis. "Once you walked on a remote surface. It's time to remember. Please, seat yourselves comfortably, close your eyes, and ... let yourself go. Welcome to the Moon. Do you remember? Can you see the electric blue shining on the horizon? That's our home, Earth. Now this memory will never leave you" (Romero 2019).

And in 2022, the film *A Mexican on the Moon* finished production in the state of Colima. It was based on a novel by Manuel Sánchez de la Madrid (2010), which was in turn based on the oral testimony of a domestic worker, who, upon watching a documentary on the anniversary of the Moon landings, claimed that Neil Armstrong had, in fact, been born in Zapotitlán de Vadillo, Colima. Therefore, according to local legend, the first man on the Moon was Mexican.[7]

Mexican NewSpace

Rosa Inés: Do you think Mexico can or should participate in space exploration?

Carlos: Well, we've only had contact since Neri Vela[8] went to space, and well, that's over, more than thirty years ago, 1985, but I still feel it would be worth it to be closer to that environment because, ultimately, it would be a solution to the things that we are missing here. They're researching materials on Mars, but it's really far. So is the Moon, but we could be surprised.[9]

"Everyone else is working on satellites," Juan Carlos told me in 2022. "We wanted to do something different."[10] "Besides, we're passionate about robots, not satellites." This passion, born (or at least nurtured) at the 2016 International Astronautic Congress held in Guadalajara,[11] led engineers Juan Carlos and César to create the start-up space company Dereum Labs; they were later joined by industrial designer Kaori and other colleagues with expertise in business and marketing. Dereum Lab's major project, the one that defines their brand and gets them funding, is Jaguar 1, a rover (a "cuberover" instead of a "cubesat," laughs César) designed to carry commercial payloads to the Moon's surface. A 2024 launch is planned in partnership with Airbus. I caught up with Juan Carlos, César, and Kaori at the Mexican Airspace Fair (FAMEX), held at a military airbase outside of Mexico City in April of 2023, where the outer space sector had been assigned its own pavilion for the first time. They were presenting their project at the Airbus stand. Airbus' "Roxy" module, designed to convert regolith into oxygen and metals through electrolysis, will use Jaguar 1 to move around the Moon. (One contact at the Mexican Space Agency, or AEM, joked: "Have you seen Roxy? She's just a *tamalera.*"[12]) Jaguar 1 is the face of Dereum (see Figure 7.2), but the start-up has recently expanded to offer business plans to help Mexican companies extend their services into outer space.

Argentinian astrophysicist Gustavo Medina Tanco and his students at the National Autonomous University of Mexico (UNAM)'s space instrumentation laboratory (LINX) were also displaying their Moon project at FAMEX. "The Moon is the 8th Continent," he declared, echoing many advocates of lunar exploration (Alvarez 2020, 3). He and his team plan to "conquer" the Moon for Mexico (preferably before Dereum) with Colmena, or "beehive," a group of six miniature robots designed to organise autonomously to construct solar panels. At one point, Colmena's robots were to be called *tepoztli*, more or less "small metal things" in Nahuatl.[13] As of the time of this writing, Colmena is already loaded

Figure 7.2 Jaguar-1. Photo by the author.

onto the Peregrine platform of the US company Astrobotic, which plans to launch its mission at some point during 2023. The mission has been delayed several times for various reasons. "What can we do?" Medina said to me, rhetorically. "We can't put any pressure on them because we're just a small client. Maybe if we were NASA."[14] If you pay, you can get to the Moon quickly. This trip is more like "travelling on a donkey. It will take months." While the team waits, they work on the next phases of the project: MoonWorm, which will consist of more "organic" robots, without little wheels and motors, and Moonscouter, which will deploy multiple robots that will help develop technology for Moon and asteroid mining. Asteroids are "an infinite resource," Medina told me. "And the point isn't to develop a *chuncho* but an entire economic area."[15]

The Moon wasn't originally "zoned" for resource extraction. Mexico was one of the signatories to the Moon Agreement of 1979, which established the principle that the Moon and other Celestial Bodies should be used only for peaceful purposes, that their environments should be undisturbed, and that the Moon and its resources should be considered the common heritage of humankind.[16] However, no space-faring countries have ratified the agreement. Mexico has also signed the Artemis Accords, which are non-binding bilateral and multilateral agreements between the US and other world governments that pledge support for the Artemis programme. Some have questioned whether the Moon Treaty and the Artemis Accords are congruent, given that the latter, while still emphasising the peaceful uses of outer space, are focused on the development of commercial activity and "sustainable resource extraction" by the US and its allies (Tronchetti and Liu 2021). But by signing the Accords, according to Medina, Mexico affirmed its interest in participating in the modern space industry.

The Moon and the dark sky

Rosa Inés: What was your first memory of space?

Perla: When I was little, well, you see the stars at night, no? And more towards dawn, you really see the stars clearly ... And that's when you say, well, we're not the only ones that are here, but also out there, and on the Moon, at night you see the Moon, too, and you say, "What would it be like to be out there? What would it be like to see the Earth from the Moon? What would the view be like?" That's when you start to imagine what's out there in space, and things like that.

Iván: The thing is that there's a better sky where we're from because we're in the east, toward Puebla. There the sky is much prettier, right? It's darker.[17]

Mexico has several planetariums (and the above-mentioned metro station) where the public can experience a simulated dark sky. And in many places, such as the megalopolis of Mexico City, these are the only options for seeing more than the Moon and a handful of stars.

In 2015, the UNESCO office in Mexico organised an international meeting on the theme of "The Right to Dark Skies" (ironically?) part of the activities held to commemorate the International Year of Light and Light-Based Technologies, sponsored by the United Nations General Assembly. The meeting was held at the Centre for Digital Culture in Mexico City in January 2016, only a few months before (again, ironically?) the IAC convened in Guadalajara to promote satellite and rocket launches. In 2021, the Mexican legislature approved reforms to the General Law for Ecological Balance and Environmental Protection that included protection against "intrusive light" and light pollution. Promoted both in the name of economic savings on energy expenditures (more successful) and the interests of astronomical research (less successful), this was the first federal law recognising the right to dark skies in Mexico.[18] Another activity that has emerged recently from this interest in the preservation of dark skies is astrotourism.

The state of Hidalgo borders Mexico City, whose light contamination, according to environmental engineer Héctor Solano Lamphar, is at the same level as Hong Kong, long considered the most "lit" city in the world (Sánchez 2016). But because of its geographical characteristics, there are regions of Hidalgo that still enjoy starry night skies. Researchers from the UNAM's Geological Institute and School of Earth Sciences have been promoting dark skies in Hidalgo for several years; at the time of this writing, they are collaborating with the community of Peña del Aire in the municipality of Huasca de Ocampo to get recognition from the International Dark Sky Association as a Dark Sky Park. In 2023, International Dark Sky Week was celebrated in the region of Huasca with the event "Voyage to the Moon in Meztitlán" ("the place of the Moon" in Nahuatl). Activities included talks about the effects of the Moon on the flights of migratory birds and the Moon in literature, the installation of a mobile planetarium, and a workshop on lunar photography.

Before the advent of eco- and astrotourism, many of Peña del Aire's inhabitants were forced to emigrate and work in the construction industry, some in Mexico City, but many as undocumented workers in the United States. Now they work as eco- and astrotourism guides. When there is no Moon, astrotourists (and anthropologists) can see the Milky Way on cloudless nights, many for the first time in their lives (see Figure 7.3). This is the kind of night that astronomers also like best. But one night that was illuminated by an almost-full moon, we took a night-time nature walk along an uneven path that led them from clifftop to creek bed. Our

Figure 7.3 Community Center, Peña del Aire. Photo by the author.

local guides paused every now and then to talk about the animals we were hearing, the stars or planets that were visible in the sky, and to tell local tales of witches and other creatures that only appeared at night. But the memory of other night-time walks is never far away: as we walk someone calls out "*La Migra!*" Some laugh, others do not.[19]

Ethnography and outer space

"How can you be an anthropologist of outer space?" I am often asked. "Wouldn't you have to go there to do research?" I do not know of any (non-fictional) anthropologists who have actually been to outer space, but I hope the first sections of this chapter have started to point the way toward some methodological choices for doing outer space ethnography that do not involve actually "being there." After all, few humans have ever "been there," but outer space is a vital part of earthly imaginaries, present in the ideas, feelings, and practices that revolve around science and technology, economy, religion and cosmology, material culture, heritage, politics, identity, art, individual and collective desires, speculations and other "futural orientations" (Bryant and Knight 2019, 16). Studying outer space inevitably requires paying attention to multiple actors, times, places, and scales; for this reason, outer space anthropology is necessarily multi-sited (Marcus 1995).

In 2017, I accompanied my high school-aged daughter to a talk at the UNAM about dark matter in which the speaker mentioned that his department had sent an experiment to the International Space Station. "The UNAM is in space," he said. I had a flashback to what I had been told years before when I was undertaking much more "traditional" fieldwork in the state of Guerrero, the birthplace of Mexico's only astronaut (see Note 6). "Guerrero is in space," one of my interlocutors said. As luck would have it, I was reading Lisa Messeri's *Placing Outer Space* (2016) at the time, and I began to think about whether there are Mexican off-Earth places and to ask extremely nebulous questions about what outer space means to Mexicans and from Mexico.

As a first step, I sent out somewhat random emails: to a few science departments at the UNAM, to Neri Vela himself, and to the AEM, of whose existence I had just become aware. The only reply I received at the time, and only a few hours after sending my queries, was from the AEM's Director of the Formation of Human Capital in the Field of Space, who invited me to his office to talk. Probably because my first contacts were from the agency's human-centred departments rather than those focused on technology or business, the AEM became a central node in what I began to explain as "a study of Mexican imaginaries of outer space." From my contacts at the agency, I learned about space instrumentation laboratories, analogous missions, regional centres for space studies, the International Astronautical Federation, and, surprisingly, space art. For the past five years, I have followed space actors from event to event, both virtual and in person, discovering new organisations, activities, and interlocutors.

My contacts at the AEM led me to a contact at the UNAM's Institute of Astronomy who organised the massive outreach festival *Noche de las Estrellas*, who led me to a contact at the Institute of Nuclear Sciences' LINX (the one sending robots to the Moon), who connected me with his collaborators at Hidalgo's governmental science and technology commission (CITNOVA), who put me in contact with the community promoting a dark sky denomination, who led me to the Mexican Dark Sky delegate at the UNAM, who led me back to *Noche de las Estrellas*. The AEM also connected me with planetariums and space artist Nahum, who connected me with his colleagues at art and technology collective Kosmica, who in turn connected me to the Institute of Nuclear Sciences, since an astrophysicist who works there had participated in one of their art projects. Another branch of the AEM network led to the group Marsarchive.org and their science-fiction workshops, which drew me to more planetariums and another network of space artists (the Mexican Space Collective), whose leader had worked with my contact at the LINX. And yet another AEM connection led to the team of participants in the first Mexican Mars analogue mission (MEX-1), which led to the start-up Dereum (the one sending rovers to the Moon), the International Space University (ISU), and the Space Generation Advisory Council (SGAC), an international organisation of students and young professionals in the space sector. And I finally got to meet Neri Vela. Looking back, some of the few aspects of my research that did not involve the AEM or its contacts, at least indirectly, were the interviews conducted by my students that came about because I was interested in hearing from people outside the "space people" network.

In the following diagram (Figure 7.4), I attempt to visualise my research space, although I have only included collectives and spaces of interaction, and not individuals or specific projects. The size of the type represents the intensity of my interactions (not the importance of the collective), names in all capital letters represent international organisations, italics indicate artistic collectives and underlined names refer to community organisations. I could have evidenced other distinctions, between government, private, and academic organisations, for example, but I decided to omit them for the sake of visual clarity.

After some ineffectual engagements with concepts such as "space community" (too positive, and used with great ambivalence by space people),

Figure 7.4 Actors in the Mexican Space Milieu.

"space ecosystem" (conceptually too confusing, although used with aspirational fervour by space people), "space networks," "communities of practice," or "epistemic communities" (only partially applicable),[20] space culture (too vague), and space microculture (too specific), I have come to understand my interlocutors as participants in a "space milieu." The term "milieu" became popular in the humanities and social sciences thanks to the work of Georges Canguilhem, whose article "The Living and its Milieu" called for paying attention to the complex and generative relations between humans and their environments (Canguilhem 2001). Milieux have been reanimated recently as analytical tools, along with related processual and indeterminate concepts, such as "worlding" and "atmosphere." Kathleen Stewart describes milieux in terms of

> ... the prismatic singularities of an actual scene of composition and decomposition forged in fractious points of contact that inspire and directly induce lines of action or simple shifts in direction or duration [...] an atmosphere with qualities, an imperative demanding a response, an objective, a pooling up that can overflow its bounds, a track on which to somehow venture out.
>
> (Stewart 2019)

Thinking with and through milieux has allowed me to orient my study towards the kinds of material and affective relationships space people have with each other, with objects and ideas, relationships that come into being through scalar tensions between the cosmos and the Earth, Mexico and the rest of the world, the past and the future. It permits me to both expand my attention beyond the minutiae of microenvironments, like agencies and laboratories and to ground the seductive but abstract philosophical questions that arise from what Brad Tabas has termed the "post planetary" condition (2022, 65). I can focus on the middle, the medium, the milieu, a perspective that draws my attention to things happening at different scales, and permits the incorporation of government functionaries, scientists, engineers, lawyers, artists, activists, and community organisers who are unlikely to participate in the same "community of practice" or have the same interests or goals regarding outer space. It gives room to both the common passions and the frictions that characterise their relationships and compose their milieu.

Interacting with my interlocutors requires following them as they move between spaces, physical and virtual, being flexible enough to let "fieldwork" emerge in the interstices of everyday life, rather than existing as a framed time and place set apart and demanding just as much "participation" as "observation." I have become part of the milieu I study, just another "space person," albeit perhaps with a less common perspective.

As in all anthropological projects, the work implies a constant oscillation between intimacy and estrangement.

My associations with the individual and collective actors in this milieu have led to extremely productive collaborations, both virtual and in-person. In my experience, participating collaboratively with diverse actors has been a way to access ethnographic information while "making myself useful." George Marcus' notion of ethnographic "para-sites" has been a helpful way to conceive of the spaces of collaboration. Marcus writes that these spaces are encounters between "counterparts" (rather than the "others" with whom anthropologists often perceive as their interlocutors), that is "coproducers of interpretations that we elicit, cajole, contest or share" (Jeevendrampillai and Fortais, Chapter 9, this volume; Marcus 2000, 2). Each encounter in a para-site "decentres the conventional ethnographer-informant relation through a para-ethnographic epistemic partnership with expert interlocutors or by involving audiences in projects of media and knowledge making" (Boyer and Marcus 2020, 13). My interlocutors have expertise in robotics, computer programming, space law, design, and artistic production. One of the benefits of working with experts from different disciplines is learning from them and generating productive dialogues around questions like "Will Mexico's first Moon mission be considered space heritage?" Or "How do dark skies impact the experience of gendered violence?" Or "Is space really for everyone?"

Some collaborations have involved conferences about the sociocultural aspects of outer space ("human factors" in the local language); this was the case for a series of encounters with the South American chapter of the SGAC in which we conceptualised what a Latin American "outer space culture" might entail. Others focus on interacting with non-specialist publics through speculative workshops that imagine diverse outer space scenarios; examples include "futuring" workshops which I helped facilitate at the AEM's headquarters in Mexico City, in Peña del Aire, and with the collective Marsarchive.org. Still others centre on practical implications of critical outer space thinking that consider gender equality, cultural diversity, and sustainability in the solution of problems, like encounters with members of the ENMICE experimental rocketry engineering group in which we discussed the social and environmental impacts of rocket launches.

Working with experts who are at least partially complicit with existing power structures comes with a set of challenges for a discipline that thinks of itself as inherently complex, critical, and on the side of the marginalised. Sometimes the notion of "collaboration" has uneasy implications in this context, and certainly, I often feel more in my comfort zone when I collaborate with artists and community organisers. That said, I have been surprised by the openness of many STEM professionals and students to social and humanistic perspectives; this has particularly been the case when working with young people, like the members of the Space Generation Advisory Council. I have found partial affinities in unlikely places. And I think that situating my research in Mexico, a country that is at the same time marginal and not marginal to space activities, has provided a more complex viewpoint on the space milieux than might have been the case for a study centred in Europe or the United States.

Final reflexions

Mexican space activities do not only revolve around the Moon, and in this chapter, I could have written about Mexican space activities that have a connection with Mars (Johnson 2023a), low earth orbit (Johnson 2023b), or any number of other cosmic points of reference. However, I felt the Moon has a particular gravitational resonance that pulls together the past and the future, Earth and the cosmos, in a manner that is conceptually conducive to introducing the Mexican space milieu and the different ways in which I have been able to engage with its actors through collaborative, mobile, and multi-sited ethnography.

This milieu has space for everyone, pun intended. It draws together artists and politicians, rocket enthusiasts, astronomers, community organisers, and everyday lunatics. It also draws together temporal registers: memory and speculation, tradition, and future technology. The winning submissions of the AEM's space art contest simultaneously imagine rabbits and rovers on the Moon. Don Quixote walked on the Moon, as did fellow fictional creation Dutalon. Mexico has a Moon rock. Nahum can make you "remember" a voyage to the Moon you never took. Former undocumented immigrants turned tourist guides, have embodied memories of moonlight intertwined with fear and hope. Mexico is indeed "the Navel of the Moon;" after all, some say Neil Armstrong was born there.

Notes

1 In the summer of 2019, I asked two of my students to interview people about their thoughts and feelings about outer space. The interviews were conducted outside of planetariums and science museums in Mexico City and in the Metro station La Raza, the site of a kind of underground planetarium or "celestial vault," through which travellers must pass. Interview between Rosa Inés Padilla, Mariel Carpio, Mónica (house cleaner), and Italia (elementary school student), Mexico City, June 2019.

2 Interview between Rosa Inés Padilla and Joshua (biochemistry student), Mexico City, June 2019.

3 This myth is taken from what may be considered the first ethnographic study of Mexico: the fifteenth-century Florentine Codex, compiled by the Franciscan Bernardino de Sahagún between 1545 and 1590.
4 Fermented agave wine.
5 Although the archaeological imagery is from the Yucatán Peninsula, the landscape and written words are allusions to Mexico's Central Valleys. "In the navel of the Moon" is one translation of "Mexico".
6 Interview between Mariel Carpio and Rey Humberto, farmer, Mexico City, June 2019.
7 See Dovey and Potts, Chapter 8, this volume, for more on the artistic possibilities of the Moon.
8 Rodolfo Neri Vela, a Mexican engineer, was chosen to represent Mexico as an astronaut aboard the shuttle *Atlantis* in 1985 when he flew as part of a mission to launch the satellite Morelos 2. To date, Neri Vela is the only astronaut to have flown representing the Mexican government, although several Mexican-American astronauts have participated in NASA missions in space.
9 Interview between Rosa Inés Padilla and Carlos, engineering professor, Mexico City, June 2019.
10 Interview with founders of Dereum Labs, Mexico City, May 2022.
11 The 2016 IAC was notable for Elon Musk's keynote speech, in which he presented his plans for colonising Mars.
12 A cylindrical pot used to steam tamales.
13 Apparently, *tepoztli* would have been too difficult to pronounce for non-Mexicans (Gustavo Medina Tanco, personal communication, April 2019).
14 Personal communication, January 2023.
15 Personal communication, August 2019. *Chuncho* is South American slang for "gadget". This chapter was written before the launch of the Colmena mission on January 8, 2024. Unfortunately, Astrobotic's lunar lander suffered problems after its launch and did not reach the Moon.
16 www.unoosa.org/pdf/gares/ARES_34_68E.pdf, consulted August 22, 2023.
17 Interview between Rosa Inés Padila, Perla, and Iván, Mexico City, June 2019.
18 A local law had already by passed in 2006 in the municipality of San Pedro Martir, Baja California, site of the National Astronomical Observatory. The law was applied statewide in 2010 to combat the problem of light contamination from the increasing use of LED lighting, as well as the light contamination resulting from the regional development of open pit mining (Avila Castro 2016, 123).
19 "*La Migra*" is a Spanish term for the US border patrol.
20 For communities of practice, see Wenger 1998; for epistemic cultures, see Knorr Cetina 1999.

References

Alvarez, Tamara. 2020. "The Eighth Continent: An Ethnography of Twenty-First Century Euro-American Plans to Settle the Moon." PhD Dissertation, New York: New School for Social Research.

Aquino, Erendira. 2016. "De México a La Luna: Una Versión Fuera de Este Mundo de Don Quijote." *Animal Político*. October 14. https://animalpolitico.com/sociedad/quijote-aniversario-espacio.

Arreola Santander, Mario. 2017. "Astronautas del Programa Apolo en México." *Hacia el espacio*. haciaelespacio.aem.gob.mx/revistadigital/articul.php?interior=523.

Avila Castro, Fernando. 2016. "La Ley Del Cielo." In *The Right to Dark Skies/El Derecho a Los Cielos Oscuros*, edited by Nuria Sanz, UNESCO, and Consejo Nacional de Ciencia y Tecnología, 123–31. México: Unesco, Oficina de México.

Boyer, Dominic, and George E. Marcus. 2020. "Introduction." In *Collaborative Anthropology Today: A Collection of Exceptions*, edited by Dominic Boyer and George E. Marcus, 1–21. Ithaca, NY: Cornell University Press.

Bryant, Rebecca, and Daniel M. Knight. 2019. *The Anthropology of the Future*. New Departures in Anthropology. Cambridge and New York: Cambridge University Press.

Canguilhem, Georges. 2001. "The Living and Its Milieu." Translated by John Savage. *Grey Room* 3 (March): 7–31. https://doi.org/10.1162/152638101300138521.

Iwaniszewski, Stanislaw. 2012. "Los Ciclos Lunares y el Calendario Maya." *Arqueología Mexicana* 19 (118): 38–42.

Johnson, Anne W. 2023a. "Mexican Mars: Narrating Spatial Futures from the Margins." In *Outer Space and Popular Culture: Influences and Interrelations, Part 3*, edited by Annette Froehlich, 1–15. Southern Space Studies. Cham: Springer International Publishing. https://doi.org/10.1007/978-3-031-25340-9.

———. 2023b. "Mexico Dreams of Satellites." In *The Routledge Handbook of Social Studies of Outer Space*, edited by Juan Francisco Salazar and Alice Gorman, 339–50. Routledge Anthropology Handbooks. London and New York: Routledge, Taylor & Francis Group.

Knorr Cetina, Karin. 1999. *Epistemic Cultures: How the Sciences Make Knowledge*. Cambridge, MA: Harvard University Press.

López-Austin, Alfredo. 1996. *The Rabbit on the Face of the Moon: Mythology in the Mesoamerican Tradition*. Salt Lake City: University of Utah Press.

Marcus, George E. 1995. "Ethnography in/of the World System: The Emergence of Multi-Sited Ethnography." *Annual Review of Anthropology* 24: 95–117.

——— (ed.). 2000. *Para-Sites: A Casebook against Cynical Reason*. Late Editions 7. Chicago, IL: University of Chicago Press.

Messeri, Lisa. 2016. *Placing Outer Space: An Earthly Ethnography of Other Worlds*. Experimental Futures. Durham, NC: Duke University Press.

Rivas, Manuel Antonio de, Carolina Depetris, and Adrián Curiel Rivera. 2009. *Sizigias y Cuadraturas Lunares Ajustadas al Meridiano de Mérida de Yucatán Por Un Anctítona o Habitador de La Luna y Dirigidas al Bachiller Don Ambrosio de Echeverría, Entonador Que Ha Sido de Kyries Funerales En La Parroquia Del Jesús de Dicha Ciudad y al Presente Profesor de Logarímica En El Pueblo de Mama de La Península de Yucatán, Para El Año Del Señor 1775: Seguido de Fragmentos Del Proceso Inquisitorial*. Vol. 1. Edited by Viajeros 1. Mérida, Yucatán: Universidad Nacional Autónoma de México.

Romero, Nahum. 2019. "Poética Orbital." Performance, May 15.
Sahagún, Bernardino. 1950. General History of the Things of New Spain: Florentine Codex. Translated by Arthur J. O. Anderson and Charles E. Dibble. Monographs of the School of American Research, no. 14, pts. 1–13. Santa Fe, NM and Salt Lake City, UT: School of American Research; University of Utah.
Sánchez de la Madrid, Manuel. 2010. *Un Mexicano en la Luna*. Colima: Diario de Colima.
Sánchez, Verenise. 2016. "México Registra Los Niveles Más Altos de Contaminación Lumínica Del Mundo." *Cienciamx* (blog). October 26. www.cienciamx.com/index.php/ciencia/ambiente/11190-mexico-registra-los-niveles-mas-altos-de-contaminacion-luminica-del-mundo#:~:text=Por%20lo%20tanto%2C%20las%20ciudades,contaminada%20de%20todo%20el%20mundo.
Stewart, Kathleen. 2019. "The Life of the Mileu." *History of Anthropology Newsletter* 43. https://histanthro.org/notes/life-of-the-milieu/.
Tabas, Brad. 2022. "Performative Infelicities on the Threshold: Why Criticism of Globalization and the Planetary Anthropocene Fuel Expansive Sustainable Development." *Ecozona*, 63–79. https://hal-ensta-bretagne.archives-ouvertes.fr/hal-03791520.
Tronchetti, Fabio, and Hao Liu. 2021. "Australia Between the Moon Agreement and the Artemis Accords." *Australian Institute of International Affairs* (blog). June 3. www.internationalaffairs.org.au/australianoutlook/australia-between-the-moon-agreement-and-the-artemis-accords/.
Wenger, Etienne. 1998. *Communities of Practice: Learning, Meaning, and Identity*. 1st ed. Cambridge: Cambridge University Press.

8 Are we still anthropologists if we go to space using only our imaginations?

Ceridwen Dovey and Rowena Potts

What happens to an ethnographic filmmaker who can no longer travel, who can barely leave their own home, let alone their neighbourhood, city, country, continent?

We both trained as social and visual anthropologists and ethnographic filmmakers (as undergraduates at Harvard University, and as postgraduate students at New York University), and have worked since then as an independent documentary filmmaker and editor (Rowena), and journalist, essayist and fiction-writer (Ceridwen). Now, in our forties, we find ourselves as primary caregivers to young children (and doing paid work outside the academy). It is no longer possible for us to make films in the ways we used to, when we were younger.

The methods and models of ethnographic filmmaking that form much of the canon in the tradition are rooted in a commitment to actuality and to encounters with the "real" through cross-cultural fieldwork. These methods and models are hard to shift or adapt. We, for example, had both been trained to "think like ethnographic filmmakers." So many of those filmmakers spent their lives travelling the globe, dipping into and out of cultures, and making films through long-term commitments to places, communities, and worlds that were not their own. The work that they made is essential and important, and we respect and admire their work; those filmmakers (Jean Rouch, David and Judith Macdougall, Tim and Patsy Asch, John Marshall, Robert Gardner, Melissa Llewelyn-Davis, Lucien Castaing-Taylor, Ilisa Barbash, Véréna Paravel, among many others) made us want to make films ourselves.

But how could we call ourselves ethnographic filmmakers if we were not holding a camera on our shoulders, squinting painfully through the lens all day, under hot sunshine and difficult circumstances? Could we still call ourselves ethnographic filmmakers if we were not making something visual from scratch, but stitching together images made by others? Could we still be anthropologists if we did fieldwork mostly online or in quiet,

DOI: 10.4324/9781003437956-8

empty museum archives, digging through troves of visual imagery of space landscapes (like the Moon), space habitats (like the International Space Station), and space memorabilia (like uneaten astronaut food), making connections and observations that still felt deeply anthropological, but did not always involve hanging out with real, live people over extended periods of time?

Much of what we had been told is core to being an anthropologist – the "being there" aspect of using your own presence, your body, as the necessary, distinctive, "live" tool of your research – was no longer available to us. Fieldwork has been the one method that anthropologists have had to set themselves apart from other social scientists, and participant-observation of other living, breathing human beings right around you assumed to be the only true anthropological way into other lives.

Yet we hope that a certain quality of paying attention is in fact what makes us anthropologists and ethnographers, whether we are paying attention to other humans in person, to images, to words, or even to invisible forces that course beneath and between us, like zeitgeists and atmospheres and moods. This is not to privilege our way of seeing as special, but in paying sustained, deep attention – in spite of living in an attention-splintered social reality – through the act of archival filmmaking or fiction-writing, we do feel as if we step briefly outside of the flow of time. We pause, research, consider, imagine, and then make something – a film, a poem – that encourages a viewer or reader to do the same.

We would like to still be able to call ourselves anthropologists, without fear of illegitimacy, though we work outside the academy and no longer remember much social theory and cannot always do fieldwork in the ways we might like to. For both of us, discovering that "anthropology" existed at the age of 18, on arriving at university, and that there was a way of being in the world, and learning about the world, and engaging with the world, that fitted exactly with what our peripatetic, ever-moving, multi-placed childhoods had already prepared us for, was life-changing. But as women who chose to have children – and could not travel with them as other (brave!) anthropologists have done – we did not anticipate that our lives could shrink down so much, that becoming rooted in one place would begin to feel inevitable. Nor could we imagine the additional impact of birthing and caring for young children during the COVID-19 pandemic, when the messiness and practical requirements of our lives suddenly – by necessity – became limited to the space afforded to us within our claustrophobic two-bedroom apartments. Can one conduct ethnographic research using a laptop set up on a cluttered kitchen table, surrounded by domestic chaos, put to frantic use during the unpredictable nap time of an infant? We would like to think so.

If our bodies had to be in one place, so be it – but surely our anthropological minds could still roam freely? And if we could only roam imaginatively – that is, imagine ourselves elsewhere – then why not roam all the way off-Earth and out into space? We made a pact to do this together. After all, as anthropologist and filmmaker Juan Francisco Salazar notes, "[I]t matters immensely from where one thinks of space, when, and why (and with whom)" (2023, xiii). We would think about space together, in stolen moments from the rest of our lives.

No anthropologist can (yet) do fieldwork off-Earth, so it is no wonder we became obsessed with thinking with – and through – space objects and space landscapes. In this way, our isolation as anthropologists stuck at home could be positively transformed into anthropologists of emotion in outer space. We could justify to ourselves our mixed methods, our dependence on archives, our ventures into speculative realms of thinking and feeling about space, and not have to explain why we weren't there (on the Moon, inside the International Space Station) in person. The prison became the prism.

There is such a fixation at the moment on the *technological means* of getting humans into space that it is easy to forget that humans have been going to space via other means – imaginative, spiritual, ancestral, cultural – since time began. Art, myth, and storytelling constitute what Angus Fletcher describes as a "narrative-emotional technology" (2022, 9) that is equally as impactful and important in shaping human history and the human future in space as any machine-based technology.

As Fletcher writes, literature (and any story-based art) does not try to tell us how to survive in the universe, but turns us "inward to grapple with the problem of surviving as ourselves" at sea in the universe (2022, 8). From the earliest times, outer space has played this role for our species, helping us to deepen the quality of our thoughts about ourselves and our place in the universe.

The ability to imagine something outside ourselves comes down to our relationship with the stars, planets, and cosmos around us – a relationship mediated not just by things/machines/rockets but by thought/poetry/art/music. Time-travel, flights of fancy, and radical imaginings of leaving Earth have had a significant impact on humanity, but those stories are usually left out of the dominant narratives about human technological, commercial, or military engagement with space.

Here is the writer Italo Calvino (1988, 7) describing how he "space travelled" in order to return to Earth with different ways of knowing:

> When the human realm seems doomed to heaviness, I feel the need to fly like Perseus into some other space. I am not talking about escaping into

> dreams or into the irrational. I mean that I have to change my approach, look at the world from a different perspective, with a different logic and with fresh methods of cognition and verification.
>
> (Calvino 1988, 7)

Writers (like Calvino), poets (like William Blake; Mina Loy; Tracy K. Smith), and musicians/filmmakers (like Sun Ra [1974]) have always known how to travel like this. Staring at the Moon may have made some people want to go there in a rocket, but it has made many more want to respond to it not via machines, but in words. As the American poet Mary Ruefle writes in a wonderful, moving essay about the history of poetry written about the Moon:

> I am convinced that the first lyric poem was written at night, and that the moon was witness to the event and that the event was witness to the moon. For me, the moon has always been the very embodiment of lyric poetry. In the West, lyric poetry begins with a woman on an island in the seventh or sixth century BC, and I say now: lyric poetry begins with a woman on an island on a moonlit night, when the moon is nearing full or just the other side of it, or on the dot.
>
> (Ruefle 2012, para. 2)

It is time for anthropologists to grant ourselves the same liberties in how we do fieldwork. When it comes to engaging with outer space, we need to draw on every part of our mental, imaginative, and emotional toolkit, and displace the "body" – our physical presence – as the only true means of observing or knowing. Digital ethnography provides one mechanism through which this is made possible (Miller 2018). Experimental modes of writing and expression in the form of ethnographically informed poetry can provide another (Maynard and Cahnmann-Taylor 2010). One of the major contributions and provocations of the emerging field of the "space humanities" is exactly this expansion beyond the "being-there-in-person" modes of fieldwork. As Salazar and Gorman acknowledge in *The Routledge Handbook of Social Studies of Outer Space*, social scientists (and perhaps especially anthropologists) can contribute not only "critical thinking" on space, but also "critical making" – which includes artistic experimentation and activism (2023, 4). It is in these "cultural borderlands" – in the liminal spaces that exist "between poetry and prose, between scholarship and art" (Maynard and Cahnmann-Taylor 2010, 4) – that we like to dwell.

As filmmakers, we have been travelling widely in space using only our imaginations for many years now. We've gone on journeys mediated and enabled by still images, archival footage, space objects in museum collections, poetry, visual artworks, and speculative fiction (see also

Jeevendrampillai and Fortais, Chapter 9, and Ojani, Chapter 3, this volume). Does this count as fieldwork? Could it count as a viable "practice of spacemaking" (Salazar and Gorman 2023, 4)?

We have found ourselves drawing on the same skills we learned as fieldworkers: how to endure the slow burn of feeling like you're going nowhere, or in circles; the way the most obscure, random, serendipitous bits and scraps end up rendering the richest insights; the moments of transcendence and beauty that suddenly emerge from long periods of boredom; or how to cope when you get a glimpse of the disturbing underbellies of the supposedly utopian and shiny human project (and futures) in space.

It helped that we'd both embraced a dual identity since completing our formal anthropological studies, as artists (filmmaker/fiction-writer) – and that our art-making had always emerged from the same impulse as our anthropology once did: to witness and watch, to metabolise things we see and learn, to experiment with voice and perspective. We value art-making because it often feels to us to be one of the last remaining safe spaces for perspective turn-taking.

Good art does not usually demand that you take a position on something; it simply asks you to immerse yourself in possibilities, letting them wash over you. (It has this in common with good anthropology.) Everything to do with activities in outer space right now is highly speculative; very little of what is currently promised will *actually* come to be possible in reality. This quality of space as a thought experiment zone, slightly outside any of the pressing realities of Earth, opens people up when they are not asked to take a side (*space settlement: for or against?*) but to think outside the planetary box, and to maybe begin to intuit that Earth is in a relationship with the rest of the universe, and always has been, but it is a relationship that plays out at a scale that is mostly beyond our grasp (see Reid, Chapter 10, this volume). This is a feeling we hoped to harness with the *Archival Futures of Outer Space Film Quartet:* four speculative, archival films that we made between 2021 and 2023.

In late 2022, we founded the Archival Futures Collective (www.archivalfutures.com) in Sydney, Australia (where we live and work) with our creative collaborators who brought the *Archival Futures of Outer Space Film Quartet* into being: sound designers, composers, audio artists, voice artists, cinematographers, and film editors. We created the Collective as a framework for ongoing collaboration, to continue to develop our practice of speculative, experimental ethnographic filmmaking.

At the core of our work is the belief that visual archives are a repository of human and more-than-human memories that can be reanimated and repurposed to say something meaningful about our present world and its possible futures. We want to make work that is imaginative and future-oriented and that relates broadly to our relationship with the environments and ecosystems that sustain us, inspire us, nurture us. We incorporate archival materials (moving and still, fragmentary and textural, discovered by happy accident and through diligent excavation) into our films, material that we source from the digital and physical collections of museums, libraries, and other cultural and scientific institutions. Our work is hybrid and always evolving. It plays at the borders of art and science, documentary and fiction.

Our interest is anthropological, but our methods are experimental. In *Experimental Ethnography: The Work of Film in the Age of Video* (1999), Catherine Russell brought ethnographic film and avant-garde cinema into radical dialogue, proposing a new category of practice. Like avant-garde film, Russell suggested, experimental ethnography requires a radical "dismantling [of] the universalist impulse of realist aesthetics" (1999, xvii). Experimental ethnographic film, therefore, tends to incorporate experimental filmmaking techniques: montage, juxtaposition, surrealism, and the use of found footage to approach and interpret other cultural worlds and spaces. Our films came into being using these techniques.

What follows are some vivid, very personal recollections of moments in the making of the *Archival Futures of Outer Space Film Quartet*. These films are designed to be viewed together, though each film (*Moonrise*, *Musca*, *Memorabilia*, *Requiem*) can also stand on its own, and each has screened at film festivals and museums in Australia and internationally. A brief description of each film is as follows:

> *Moonrise* (2021, 11 minutes) is an archival film-poem imagining the Moon addressing Earth in a monologue that acknowledges their joint history. It ends with the Moon turning away from Earth, looking out towards the universe's larger and ever-expanding webs of connection.
>
> *Musca* (2022, 5 minutes) is a reflection on the naming of constellations in the Southern Hemisphere skies, and a playful critique of how only certain humans have historically had the power to name constellations on maps and star-charts for posterity.
>
> *Memorabilia* (2023, 17 minutes) uses imagery of real space memorabilia in museum space collections to pose questions about the human tendency to attach emotional and spiritual significance to "flown" objects that are in some way touched by the "magic" of outer space.
>
> *Requiem* (2023, 16 minutes) imagines the final astronauts on the International Space Station (ISS) bidding farewell to their beloved habitat through a sonnet cycle, before it is deorbited and parts of it crash into a remote area of the South Pacific Ocean, an event currently planned for around 2030 (NASA 2022). We arranged for the sonnet

cycle to be translated into several languages and voice-performed for the film by real-life astronauts who have spent time on the ISS or Mir space stations, including Cady Coleman (USA), Paolo Nespoli (Italy), Dorin Prunariu (Romanian astronaut within the Soviet space programme), Soyeon Yi (South Korea), and Claudie Haigneré (France).

Moment 1: Participant-observation of the Moon

We are in a sound booth in a studio that has kindly let us be there for free for an hour so we can record the voice of the Moon for our film *Moonrise*. Or, more accurately: the twinned, dual-gender *voices* of the Moon. Due to lack of budget, Rowena has gamely agreed to be the female voice of the Moon. Our friend Sepehr Jamshidi Fard – a radio producer and storyteller – is going to be the male voice of the Moon.

Rowena and Sepehr have been practising over Zoom for weeks, reciting the poem – a monologue from the perspective of the Moon, addressed to Earth – over and over, trying to get their voices and pace in sync. Now they stand opposite each other on either side of a standing microphone. Our sound designer, Annie Breslin, is sitting inside the booth, coaxing and guiding them like a conductor. They speak the words of the poem slowly, simultaneously (below are the first two stanzas of the much longer poem):

Moonrise
I am full tonight, so full my light might overflow
and I'm in the mood to commune.
If you would pause in your dizzying spin –
come, rest a while, as the celestial fires encircle us …
let my umbral shadow fall gently upon you. It's time
for you and I – I and Thou – who've spent aeons side by side
to reckon with who we've been, and who we might become.

I'll start, if you like … it's a gift, mostly, to gaze at you:
Earth – my companion, the one to whom I am in thrall.
Though it's painful, sometimes, to live beside such luminous proof
of life,
unable to un-see your abundance, the way you teem
so boastfully with blues and greens
while here I wax and wane for you, my grey surface changing
so slowly
I must seem almost the same as when we were first made.

This poem – which became the script for *Moonrise* – emerged from several years of thinking and writing about the Moon as a wilderness landscape that deserves to be protected and conserved (Dovey 2021). Is it possible to create "nature-writing" about outer space? Like ethnographic fieldwork, certain types of nature-writing tend to depend on in-person witnessing, but we wondered if it could be done remotely instead, and propelled not by a nostalgic, elegiac, backwards-looking gaze at all that has been lost, but by anticipating all that may one day be lost in the future.

Thinking about space-nature means pivoting to consider our responsibilities to nature in the future. This is a reverse longing, in a sense, for the space landscapes we hope and dream will survive any encounter with humans. It is akin to "solastalgia," a relatively recent term used to refer to the human experience of emotional distress and sadness generated by the witnessing of environmental degradation. While we may not yet have experienced what nature writer James Bradley calls the "ecological grief" that inspires much contemporary nature-writing on Earth, we need to consider the possibility that any human future in space is also the beginning of grief over what humans will inevitably try to take from those places (Bradley 2020, n.p.).

The posthuman, interspecies, multispecies, and more-than-human modes within nature-writing and the wider environmental humanities and arts (Celemajer 2021; Feral Atlas 2021; Singer 2016; van Dooren, Kirksey and Münster 2016; Tsing 2015) already allow for freely and creatively thinking with or through or against or amongst non-human entities such as landscapes, places, non-human species, objects, and things. After all, places are "co-constituted in processes of overlapping and entangled 'storying' in which different participants may have very different ideas about where we have come from and where we are going" (van Dooren and Rose 2012, 3).

This is true for space-places, too – even ones that most humans will not be able to set foot on themselves for the foreseeable future. The untangling of those knotted stories about the Moon can only (for now) be done from a distance, yet this distance can also engender emotional proximity in its own way. As director and editor of *Moonrise*, Rowena sought to vivify Ceridwen's poetic "storying" of the Moon's perspective through montage. She gleaned imagery from virtual fields of digital visual material, picking through the archival collections of museums, universities, research institutes, cultural organisations, and space agencies.

Her searching, sorting, and sifting process resulted in multiple folders of downloaded imagery, and a massive spreadsheet with descriptions of each item for reference throughout the editing process. The diverse representations of the Moon that she collected came from archives and digital collections held by the Library of Congress, the New York Public Library, the Metropolitan Museum of Art, the British Museum, the Getty Museum, the Rijks Museum, the Prelinger Archives, NASA's

Data Visualisation Studio, NASA's Lunar Reconnaissance Orbiter, the Goddard Space Flight Centre Archives, the European Space Agency, the US Geological Survey, Bryn Mawr University Special Collections, and Australia's National Film and Sound Archive, as well as several private collections.

This material included still and moving imagery, black-and-white as well as colour, contemporary, historical, cultural, kitsch, scientific, artistic, virtual, real. Taken together, this imagery provided a foundation for the method by which the human relationship to the Moon – as a place, a landscape, an object, a story, an idea, a hope, a Self, an Other – could begin to be untangled.

But what also happened was this: through this act of close looking at all these images of the Moon, a celestial object mediated in so many ways across time and space, we both began to build an intimate and emotional relationship with the Moon.

How can one look at so many thousands of photographs, models, sketches, paintings, woodblocks, maps, prints, films – human visualisations (see Figure 8.1), human objectifications – of the Moon (and its pockmarked yet pristine lunar surface) and not begin to feel grief about its future, if that future is determined by humans for whom it represents only an exploitable resource? The layering of the recorded Moon monologue over, under,

Figure 8.1 Still from *Moonrise*. From: *Sky: A Film Lesson in Nature Studies*, 1928. Courtesy of Prelinger Archives.

and through this imagery felt like an act of close listening, symbolically linking us with so many other humans, past and present, in feeling a deep reverence for the Moon (Slaughter 1902). Our creative work with digital archives of the Moon constitutes its own form of nature-writing, as well as being a kind of experimental ethnography. Our films are filled with fieldnotes from our archival journeys.

Moment 2: Obsessive feelings about space food

It's a Tuesday in winter, the post-lunch slump. We are at a science museum in Sydney, where we've been given permission to explore their space memorabilia collection and make a film. Rowena's baby Iris, who is our honorary research assistant, has stared at every space object we inspected that morning down in the basement with open wonder, and is now taking a nap in her pram beside us in the Research Fellows' Room.

We are going through the museum's Blue Books – blue folders stuffed with old accession paperwork about individual objects. We are both obsessed with these old folders of paper, analogue methods that are soon to vanish into digital: the papers are tagged to be scanned and shredded. We are allowed to touch these folders, whereas the space objects themselves are out of our reach, only to be handled by the blue-gloved hands of curators and registrars while we stand and gaze at them longingly: rows of uneaten items of astronaut space food, Soviet and American (cabbage puree in a tube, a tin labelled "tiny sausages," a vacuum-packed cheese sandwich).

Going through the Blue Books, our attention snags on the same things. Faxes from the Florida space dealer who sold the museum many of these objects, often with annotated, handwritten notes affirming the authenticity of a particular space object, and that it had been "flown" to space and back. This word, FLOWN, is all over the paperwork in capitals, like a mantra. Yet beneath the outward confidence of the dealer's assertions simmers real anxiety: for how can it ever really be proven that this cube of compressed peanuts flew to space and back?

If it didn't fly to space and back, it is not quite as magical (or valuable) as the cubes of compressed peanuts that remained uneaten in the astronaut food storage room at NASA, and then were maybe (illegally) sold on the black market. NASA's rules as to whether or not this kind of space memorabilia can be sold by ex-astronauts and ex-employees have tended to be strict in the past, though they have relaxed them more recently (Pearlman 2012).

We can see in the paperwork what the museum paid at long-ago space auctions to purchase other bits of "flown" space food (see Figure 8.2). We begin to seek out space auction videos online, and keep an eye on the

Figure 8.2 Cosmonaut space meal manufactured by Soviet/Russian Academy of Sciences (Moscow), Russia, 1993. Image by Rowena Potts, courtesy of Powerhouse Museum and object donor. Gift of Rees Hughes, 1994.

promo videos for upcoming space auctions at Christie's and Sotheby's. These items can go for quite a lot, and yet it can all feel a bit silly, like a game or a joke. Sometimes we laugh out loud: the first eating utensils in space are to be auctioned, or a razor someone used to shave their chin on the ISS, or an eye mask an astronaut used to try to sleep. It's either the high-prestige hardware (a rover, a lander, a probe) or the intimate care items that touched astronaut bodies (food, toiletries, sleeping bags) that seem to go for the most money.

For much of anthropology's history, the study of material culture was made possible by the removal of objects from the field (often under troubling ethical circumstances). Stories and false assumptions, guesses and conjecture about these objects circulated alongside the objects themselves. "Flown" space objects are envoys from the field too, ones that have returned from out there, brought back down to the ground and enthusiastically collected by humans who understand that their symbolic value might translate into financial value (see Osbourne, Chapter 5, this volume, for a contrasting account of how non-human-made extraterrestrial objects are displayed and valued).

Take the Apollo 14 Moon tree seeds: a packet of seeds that flew to the Moon and back on the Apollo 14 mission, and were then distributed randomly across America to grow into trees, a "living legacy" of the connection between Earth and Moon (Koren 2022, n.p.). Simply by going to space – and returning – these seeds are transformed into holy relics, as are all the other banal or mundane objects that made the same journey up and down:

> All the trinkets and tchotchkes that the Apollo astronauts took with them in their personal canvas bags are cool for this reason, bestowed with a magical sheen the second they were returned to Earth—space souvenirs.
>
> (Koren 2022, n.p.)

The philosopher Jane Bennett defines an "enchanted" object as one that leaves humans "transfixed, spellbound," and "struck and shaken by the extraordinary that lives amid the familiar and the everyday" (2001, 4–5). Yet we began to see that not all space memorabilia is created equal. The emotions that space objects inspire in humans can be radically variable. The status of space objects as enchanted is also never stable: within even just one object's story-span, this status can be reversed (from enchanted object to discarded object), or can shade quickly from enchantment to disenchantment, even into fear or horror.

Why do some space objects inspire feelings of wonder and awe, while others elicit more ambivalent emotions? Which objects do we respect as our dignified proxies in space, and which do we resent for reminding us that the human body is not made to survive in space the way inanimate objects can? As we spent time with the objects that had *returned* to Earth from space – whether by choice or force or gravity or accident – we realised that they are treated reverentially in part because they *came back to us* (see also Tereshin and Sivkov, Chapter 6, this volume).

This stands in stark contrast to most human-made space objects – satellites, probes, Starman in a cherry-red convertible, the Voyager spacecraft, Mars rovers – which no longer really need us once they're out there. They thrive in places we would fail. We become hopelessly dependent on them to report back on what they see and learn (and yet we sometimes distrust their reliability as witnesses). They persist in space, inorganic and almost immortal, as our organic bodies age and die on Earth.

The space objects that refuse to return to us, or the ones that only return in pieces, mock and confuse our hierarchies of worthy life-forms by being neither alive nor dead. We rejoice in their discoveries, follow every step of their adventures as if they are pilgrims, and mourn their loss even if we are the ones responsible for burning them up in the atmosphere or crashing them into the ocean. Some of them keep journeying even long after they are no longer responsive to us, instruments switched off, gone dark. Every launch of a space object is a birth and a death. For the object, it is the start of a grand adventure, one that – if the object is lucky – may take place outside human surveillance, under cover of the darkness of space.

Certain space objects are disobedient to humans in ways we both applaud and fear, an exhibition curated by the Victoria and Albert Museum in London in 2014) (Flood and Grindon 2014). What interests us is the subversive accidental disobedience that many fairly ordinary space objects achieve in space, and the "taming" or "silencing" of space memorabilia that happens in a museum collection. As the *Disobedient Objects* catalogue notes, "There is always a danger that in placing an object in a museum, you silence it" (Flood and Grindon 2014, 130).

When human-made objects are absorbed by outer space and do not return, they gain an agency of a kind they never had on Earth, freed from the bounds of gravity. They travel and circulate, even if circuitously – and that movement, whether outwards and outwards (like the Voyager spacecraft) or around and around in orbit (like the International Space Station) is what animates them into approximating being alive. Humans may still notionally control them – as we do the ISS – but often that control is lost over time. The twin Voyager spacecraft's communication and scientific instruments have had to be turned off, one by one, the further away they get from Earth. They transform into rogue objects, invisible to us in any spectrum, out of our communicative range.

Even Starman, though he is screwed into place in his cherry-red convertible, is no longer responsive to SpaceX's commands or control. Maybe he's having a wonderful time, or maybe he is furious at being abandoned to a terrible fate (Ceridwen further explores the inner lives of space objects like Starman, the ISS, the Voyager spacecraft, and the first statue left on the Moon in her book of short stories *Only the Astronauts* [Dovey 2024]).

But the "flown" space food in the museum came back down to Earth. In return for its loyalty to humans, we have promised to look after it in climate-controlled vaults, and to speak of it in hushed tones, and to light it carefully when it is on display, and to only handle it wearing blue plastic gloves. We are grateful to it for staying with us, for choosing Earth over the universe, even though our initial plan was to make it disappear by eating it. Now, of course, we would never dare to look at the "flown" Soviet sugar space biscuit hungrily. Its immortality is guaranteed.

At the end of our film *Memorabilia*, the daughter who has inherited her father's space food collection commits an act of sacrilege, and decides that she *will* eat that "flown" Soviet sugar space biscuit. She eats it because she wants to become worthy of the same kind of reverence her father gave to his space memorabilia. To film this scene, we found a ginger biscuit that looked almost identical at the local supermarket, and (again for lack of budget) Ceridwen acted as the daughter in the scene, pouring hot tea, unwrapping the biscuit, and eating it to the last crumb. Sometimes people gasp when they watch that scene.

Once the biscuit has been eaten, these words appear on the screen:

> *Inside this ordinary daughter's belly is a partially digested Soviet sugar biscuit that once flew to outer space and back. I hereby certify her thus to be worth careful keeping for posterity, extremely valuable, and touched by magic.*

Thanks to this extended period of time we spent in the museum as participant-observers – peppering the curators with questions about the value attached to a flown thermal tile fragment or a sachet of preserved cherry juice, watching registrars carefully handle and move objects from shelf to trolley to digitisation station and back, scouring the documentation of decades of space food collections in piles of Blue Books – we stumbled upon the emotional power of "flown" space objects as a concept worthy of playful interrogation in our film. Herein lies the value of ethnography conducted in archives (with a baby in tow).

Moment 3: Bringing the ISS home

It's 8:55 p.m. on a Thursday evening in Sydney, and Rowena is sitting at her desk, breastfeeding her five-month-old daughter Iris and chatting on Zoom with her close friend, sound designer Annie Breslin. They are waiting for Dr Claudie Haigneré, the first French woman to go to space, to arrive online for a scheduled remote recording session for our film *Requiem*. Dr Haigneré is a former astronaut with the Centre National d'Études Spatiales (1985–99) and the European Space Agency (1999–2002). She also has the unique distinction of having spent time on Mir (in 1993), as well as the International Space Station (in 2001) (see Figure 8.3).

When Dr Haigneré appears online, her Zoom background is of the famous ISS cupola window. She is beautiful. Her hair is white-grey, tucked neatly behind her ears, and she greets the team of three (Rowena, Annie, and baby Iris) warmly. We have asked Dr Haigneré to read a sonnet written by Ceridwen, from the perspective of an imaginary future astronaut onboard the ISS, bidding farewell to the vessel on the eve of its destruction. Incredibly, she is game.

What does this mean? It means that a highly trained scientist-astronaut with real-life experience of what it is like to eat, sleep, dream, work, float, laugh, worry, cry, *be* – on more than one space station! – is willing to leave

Figure 8.3 View of ISS Port-Zenith from Departing Soyuz TMA-20 Spacecraft Complete ISS stack including Space Shuttle *Endeavour* and Automated Transfer Vehicle Kepler International Space Station – ISS Low Earth Orbit, Space © Photo by Paolo Nespoli, courtesy of NASA and ASI.

the realm of the actual and enter a space of imagination. She's prepared to do this to help people who will never spend time on the ISS begin to grasp how much meaning this home in space has always had for its inhabitants.

We begin recording. Dr Haigneré recites the sonnet: a rich, emotion-infused performance. Annie offers some guidance and notes: "What about a longer pause between these two lines?" "Can you try a slightly different inflection here, Claudie?" We listen, amazed at the experience of sharing virtual space with this amazing woman, watching as she reads lines of poetry with care and precision for an experimental film about the ISS which is only just starting to take shape.

Astronauts are not often given the opportunity to express emotion as it relates to their experiences in space. But these sonnets speak directly to Dr Haigneré and her experience of life on space stations. This requiem for the ISS also feels to her like a requiem for Mir, she tells us. The emotions captured in the sonnets remind her of how she felt when she departed Mir, which was brought out of its orbit and destroyed in 2001, and now lies in pieces in the Pacific Ocean, where the ISS may one day join it.

Is this fieldwork? Not exactly. And yet! And yet. In some ways we were briefly up there with her, reading poetry together in the cupola of the ISS, and reflecting on her time on two different space stations and the ten years she spent living in Star City, Russia.

Over several months, as we remotely recorded sonnet after sonnet with each of the six international astronauts who participated in *Requiem*, we were able to gain tangible glimpses of a field site we would never visit. In one conversation, we learned of the grief an astronaut felt after leaving the ISS for the last time. "I grieved when I left," she said. "I really grieved." Another described her happiness and laughter while chasing Jelly Belly jellybeans that spilled out of a packet in the microgravity environment. These glimpses of ordinary, daily life on the ISS illuminated and informed our film, adding warp and weft to an otherwise speculative overall fabric. By engaging with real astronauts in this way, we did manage to simulate the anthropologically fetishised activity of "being there" through ethnographic encounter, mediated by performances of poetry.

As artist-anthropologist-filmmakers, we believe that any imagined human future in space should be animated not by technological or economic determinants alone, but by a wide array of existential, ethical, ecological, and poetic possibilities. Through our films, we gently observe and probe the strange belief in human perfectionism in space. We are interested in exploring the opposite idea: anti-perfectionism in space. How we fail in space, how we go to space and sometimes become our worst selves. How we are not made to survive in space, how our bodies betray us up there, and how that makes us feel. And how space can be a zone of death, discomfort, terror, and ambivalence, as well as a zone of wonder, awe, and beauty.

We have been moved to watch some viewers of our films come into the cinema charged up with certain ideas or assumptions (for instance, that space ethics or space environmentalism is dumb, that rockets are awesome and humans should settle anywhere they please in the universe, that the Moon is nothing but a dead rock and it is ridiculous to think of it as a nature-place), and then soften and breathe out and simply sit in a space of feeling and not knowing. We have seen guards come down, and minds open up. This was always our hope: that our films might pull together some of the loose threads of an anthropologist's unique way of being in the world, and shift people's epistemic certainty – just a tiny bit, just for a moment.

References

Bennett, J. 2001. *The Enchantment of Modern Life*. Princeton University Press.

Bradley, J. 2020. "New Nature." Panel discussion, State Library of NSW & The Sydney Review of Books, Sydney, May 23.

Calvino, I. 1988. *Six Memos for the Next Millennium*. Harvard University Press.

Celemajer, D. 2021. *Summertime: Reflections on a Vanishing Future*. Penguin Random House.

Dovey, C. 2021. "Our Sister Satellite." In *Kinship: Belonging in a World of Relations,* Vol. 1: Planet, edited by Gavin Van Horn, Robin Wall Kimmerer & John Hausdoerffer. Center for Humans and Nature Press.
Dovey, C. 2024. *Only the Astronauts*. Penguin Random House Australia.
Feral Atlas. 2021. Curated and edited by Anna L. Tsing, Jennifer Deger, Alder Keleman Saxena & Feifei Zhou. Stanford University. https://feralatlas.supdigital.org/world/acceleration?cd=true&rr=true&cdex=true [Accessed May 15, 2023]
Fletcher, A. 2022. *Wonderworks: Literary Invention and the Science of Stories*. Simon & Schuster.
Flood, C. & Grindon, G. (Eds.). 2014. *Disobedient Objects*. V&A Publishing.
Koren, M. 2022. "We Almost Forgot About the Moon Trees." *The Atlantic*, January 31 www.theatlantic.com/science/archive/2022/01/nasa-moon-trees-apollo/621395/
Maynard, K. & Cahnmann-Taylor, M. 2010. "Anthropology at the Edge of Words: Where Poetry and Ethnography Meet." *Anthropology and Humanism* 35(1): 2–19.
Miller, Daniel. [2018] 2023. "Digital anthropology". In *The Open Encyclopedia of Anthropology*, edited by Felix Stein. Facsimile of the first edition in *The Cambridge Encyclopedia of Anthropology*. http://doi.org/10.29164/18digital
NASA. 2022. "FAQ: The International Space Station 2022 Transition Plan." Published February 12 www.nasa.gov/feature/faq-the-international-space-station-2022-transition-plan [Accessed May 15, 2023]
Pearlman, R.Z. 2012. "Astronauts Can Keep (Or Sell) Their Space Artifacts, New Law Says." *space.com*, September 28 www.space.com/17787-nasa-astronauts-space-artifacts-souvenirs-law.html
Ruefle, Mary. 2012. "Poetry and the Moon." https://poets.org/text/poetry-and-moon
Russell, C. 1999. *Experimental Ethnography: The Work of Film in the Age of Video*. Duke University Press.
Salazar, J.F. 2023. "Preface." In *The Routledge Handbook of Social Studies of Outer Space*, edited by Juan Francisco Salazar & Alice Gorman. Routledge.
Salazar, J.F. & Gorman, A. (Eds.). 2023. *The Routledge Handbook of Social Studies of Outer Space*. Routledge.
Singer, H. 2016. "Writing the Fleischgeist." *Animal Studies Journal* 5(2): 183–201, https://ro.uow.edu.au/asj/vol5/iss2/9
Slaughter, J.W. 1902. "The Moon in Childhood and Folklore." *American Journal of Psychology* 13(2): 294–318.
Sun Ra. 1974. *Space is the Place* [Motion picture]. 85 minutes. Plexifilm.
Tsing, A.L. 2015. *The Mushroom at the End of the World: On the Possibility of Life in Capitalist Ruins*. Princeton University Press.
van Dooren, T., Kirksey, E., Münster, U. 2016. "Multispecies Studies: Cultivating Arts of Attentiveness." *Environmental Humanities* 8(1): 1–23, https://doi.org/10.1215/22011919-3527695
van Dooren, T., Rose, D.B. 2012. "Storied-Places in a Multispecies City." *Humanimalia* 3(2): 1–27, https://doi.org/10.52537/humanimalia.10046

9 Terraforming a field site

Reflections on crafting knowledge on Mars

David (Jeeva) Jeevendrampillai and Sarah Fortais

In the spring of 2022, twenty nineteen interdisciplinary researchers inhabited a remote Scottish island for five days. Whilst there, five analogue astronauts (researchers training in a simulation of an astronaut mission) were tasked with surveying the island. They were told it was a terraformed Martian landscape (the rest of the crew affectionately called them the Tartan Martians). *During this time, they encountered several distressed crew members from previous missions.* Over three eight-hour simulated scenarios, the astronauts had to rescue and treat the troubled crew, provide healthcare, and bring the crew to safety whilst carrying out their assigned mission of land surveying, to the best of their ability. Other 'non-astronaut' researchers measured the astronauts' changing heart rates, observed their ability to make decisions under conditions of stress, and assessed the strategies, social relations, and engagements with landscapes that emerged throughout the analogue astronaut simulation.

Both authors of this chapter were part of the non-astronaut crew. The non-astronaut crew was comprised of researchers interested in astro-pharmacy, heart monitoring equipment, and crew dynamics, all of whom needed this mission simulation to test their work. Other mission crew included trained mountain leaders, support staff, a team medic, and ourselves. The mission, organised by University College London (UCL) researcher Myles Harris, was supported, in part, by UCL's Centre of Outer Space Studies (COSS). Jeeva, the founder and director of COSS, was on the mission to help develop the analogue, observe crew dynamics, and support the work of Sarah, the artist-in-residence at COSS. As an artist practitioner, she aimed to produce several site-specific artworks, create a visual identity for the mission, and increase mission fidelity of the simulation. One of the aims of her artist residency at COSS was to think through the practice of space science differently. An artist, through their practice, can bring different elements, such as aesthetics, bodies, materials, and objects, into relation in a way that opens new avenues of thinking within

DOI: 10.4324/9781003437956-9

a critical approach to space science. We aimed to use artistic practice less as a reflection of space science or as a medium of public outreach but rather as a method of ethnographic engagement in itself (cf. Sansi 2020). We wanted to use art to illuminate certain aspects of the mission, such as astronauts' engagement with the landscape, the act of imagining being on *Mars*, and the role of non-humans. In recounting and reflecting on this experience, we chose a writing style that places the reader within the act of the mission itself by simulating the vacillation between different modes of imagining involved in a simulated mission to *Mars*. This chapter presents a form of ethnographic method and write-up that emerges from an anthropologist and an artist working alongside each other to think through the act of terraforming Scotland into a *Martian landscape* via bricolage, play, and performance.

Earth/Mars vacillation

As a boat became a *shuttle*, the island's resident deer ticks became a vector of *radiation*, and rain became a *Martian sandstorm* the astronauts and mission crew vacillated between Earth and *Mars*. Making and telling ghost stories and being playful as we encountered the landscape, found materials, and creatures allowed us to perform and inhabit *Mars* in our own particular way. Earth (a Scottish island) merged with our Martian simulation to create a sense of being in someplace unique to us in this moment. Further, this 'sinking into' *Mars* enabled the crew to simulate conditions of extraterrestrial places, where critical healthcare decision-making must be made under extreme, remote, and resource-scarce conditions. Whilst performative work increases the effectiveness of the analogue as a research method, it was noted that the crew would leave their simulated roles and places in a Martian landscape to return temporarily to Earth. Their characters would slip from Martian crew roles to their Earthly selves, points of reference would move from 'Mars' being cold to 'Scotland' being cold, and they would slip in how they talked of the future as they shifted from discussing 'how do we get back to Earth?' to 'which train are you catching to London?' This enabled them to maintain a degree of social relations at times of mission stress by switching from 'Mars talk' to 'Earth talk.' The astronauts and other researchers, who would also play with Mars building in their analogue activity, were reaching forward to possible future worlds. They were looking to develop new medical procedures, but in doing so were also thinking about their relationship to place, other species, land, territory, and their bodies differently. As such, this chapter is situated in an in-between fieldsite, and it takes a paraethnographic approach (Lambert-Beatty 2009). That is, it plays with narrative to situate the reader in the landscapes multiple (cf. Mol 2002) of Scotland and *Mars*, thereby taking the reader on a journey into paraethnographic other worlds, akin to how Annemarie Mol presents multiple texts in her book *The Body Multiple*, an ethnography of the diagnosis and treatment of atherosclerosis. One text is ethnographic data, and alongside it she presents her academic analysis. Mol shows how the disease is made to cohere through the multiple perspectives of those making images, researching, diagnosing, treating, and suffering the disease. Taking inspiration from Mol's work, we also present multiple texts, not of ethnography and analysis, but of multiple ethnographic registers. We argue that this ethnography is not of either Earth or *Mars*, but of a place that is of both. Both are real, and both are imagined. Our interlocutors interact, play, and tell stories with the landscape around them to build worlds. We invite you to inhabit the vacillatory position as our text moves between different worlds.

Borrowing from Carrie Lambert-Beatty (2009), a 'parafiction' is a partly real and partly fiction narrative. Here we would prefer to say that our paraethnography is partly experienced and partly performed as both the 'real' and the 'fictive,' as the 'sensed landscape' and 'imagined one' work together to collapse distinctions and create the realness of the Earth/Mars experience. Parafictions, according to Lambert-Beatty, are used to show how other ways of being might be possible. Using linguistic theory from Jacques Derrida (1988 [1971]) and J.L. Austin (1975), Lambert-Beatty describes how a 'parafictioner' may disrupt established lines of meaning between objects and the social worlds they are embedded within. Through performative and coercive acts, a parafictioner can produce and manage new meanings and build new worlds. Parafictions, writes Lambert-Beatty, 'intervene in what Jacques Rancière calls the distribution of the sensible: the system of inclusions and exclusions that determine what can be sensed; the literally common sense about what can be said, thought, seen, felt, and who can say, think, see, and feel it.' Hence, 'a new distribution of the sensible has, at least temporarily, been brought into being' (Lambert-Beatty 2009, 64). Through temporarily inhabiting other worlds, the astronauts bring about a position of not quite being on Earth nor Mars but productively vacillate between states of being and desire. They act out the future in a sensorial, corporeal manner, using their bodies, the landscape, the nonhuman, and their imagination to enhance a vision of a possible future in another world. They work their bodies over the land, they sweat, remove ticks, imagine radiation, see rain as a sandstorm, and so on. Their phenomenological experience is neither of Earth nor of *Mars*, but of the place in-between.

Our ethnography is para in another sense. George Marcus outlines how ethnographic knowledge often emerges less from an observation of the field but through collaborative practice with one's interlocutors. He calls this space of collaboration that emerges from encounters *between* 'counterparts' rather than *with* interlocutor 'others' – a 'para-site.' As Johnson (Chapter 7, this volume) also notes, this positions ethnographic interlocutors as 'coproducers of interpretations that we elicit, cajole, contest or share' (Marcus 2000:2). This chapter was written in this spirit.

Just as the astronauts did in the field, the narrative of this chapter's ethnography vacillates between ethnography as analysis (observation) and ethnography as performed (participation). The purpose of this experiment in narrative style is to draw the reader's attention to the nature of the vacillating fieldsite. Following Gassan Hage's provocation that 'vacillation is not just a movement between various states of being; rather, it is a state of being in itself' (Hage 2017, 202; Hage 2009), we aim to situate the reader there. Whilst talking about the experiences amongst the Lebanese diaspora and migrants generally, Hage argues that migration is more than moving from one place to another – rather it is about chasing a desire for something else, to improve one's life. In this sense, our astronauts, and those invested in making human life off-Earth, traverse the Scottish island in a state of vacillation between the imaginings of being a Martian explorer dealing with rocks, sand, and radiation, and the reality of being on Earth, with its mud, rain, and ticks. Here the astronauts are projecting a desire to sculpt off-Earth capacities, to make a future where going to such places is made possible. They do so through the acts of testing, simulating, and performing of the processes and procedures that such a move to *Mars* would require.

Bricolage

To describe this process of making and sculpting a Mars-based future, and further our ethnographic and artistic methodology, we lean on Claude Lévi-Strauss's notion of the bricoleur. The etymology of the word bricolage relates to the French verb *bricoler*, which originated in the context of 'ball games, billiards, hunting, shooting, or riding' (Lévi-Strauss, 1962, 1). *Bricoler* referred to some extraneous movement: a ball rebounding, a dog straying, or a horse swerving from its direct course to avoid an obstacle. *Bricoler* therefore indicates the physical act of deviating from a direct route, as a direct response to one's environment. The term *bricoleur* was, then, used by Lévi-Strauss in 1962 to describe a methodology of creating and acquiring knowledge that is opposite to engineering (1962, 10–12). Both the bricoleur and the engineer employ a variety of methods to produce knowledge/objects. However, an engineer does so by first creating/acquiring tools for specialised purposes, while a bricoleur improvises and uses non-specialised tools for a variety of purposes, rebounding and swerving to achieve their aims.

Lévi-Strauss stresses that a bricoleur works with their hands like a craftsperson and that a bricoleur is always making do with whatever is at hand to make knowledge. Whilst the bricoleur and engineer are both goal-oriented, the bricoleur's primary interest lies within the process: the manipulation of materials. It is the openness to the processes of knowledge production that we wish to emphasise here. There were many types of bricoleurs at work in this setting. Firstly, the astronauts and crew on the mission were getting by with that which was at hand. The bricoleur, as someone who appropriates, repurposes, and reuses in situ resources is an apt description for an imagined Martian astronaut. The astronauts were taking what they found in situ and repurposing it with a little more intent than that which might be said of a bricoleur. They, along with the wider mission crew, were also doing the work of performing being on Mars by using what they had at hand alongside the natural affordances of the terrain. Evidence from their WhatsApp communications showed that they situated most of their environment and props into a Martian landscape describing a *drop zone* and *mission base* rather than referring to a 'hut on the beach' or 'tent.'

Secondly, the artist, Sarah, a self-defined bricoleur, was gathering materials from the island to think (and feel) through the forms of relations that emerged between people, places, animals, and materials whilst on a remote Scottish island imagining being on Mars. Sarah followed the astronauts on mission simulations; she watched their body gestures, and their movements through the landscape. Like the rest of the crew, she didn't interact with the astronauts during the simulations. However, she did co-develop a specifically designed mission patch and flag with them before the mission and was heavily involved in the debrief and post-mission work through her artworks. In her previous work (Fortais 2018), Sarah has used the term 'adhocism,' following Charles Jencks and Nathen Silver (2013), when describing her methodology to produce artworks. Adhocism, whilst similar to bricolage in terms of its use of available materials, also denotes a principle of action having speed or economy, purpose, or utility, and which lives on the edge of respectability. It involves using an available system in a new way to solve a problem quickly and efficiently. Jencks and Silver note that 'Adhocism rarely presents anything new in the sense of a discovery' since, 'the purpose is to help solve a problem, or to change contexts and make it a non-problem' (Jencks and Silver 2013, 115). The novelty of adhocism comes from its improvised methods.

Sarah also uses the term 'sensibility' to refer to her process of selecting materials found in the environment she works in. Here the landscape is sensed, not only in the way described by phenomenologists of landscape (Tilley 1994; Tilley and Cameron-Daum 2017), whereby people make sense of the world through their bodies as they engage with sights, sounds, tastes, and feel. Here the artist senses the landscape with a further purpose to produce artworks. There is a conscious process of crafting the feeling of being in a particular place. This is more intentional and agentive than is often theorised by the classic phenomenological approaches to people's engagement with the landscape (cf. Rose, Degan, and Basdas 2010). The artworks Sarah produced from this mission were collectively displayed in an exhibition called 'Martian Picnic' during the 'Explore Festival' at the Royal Geographical Society in November 2022. The artworks communicate beyond text. They allow the viewer to gain a sense of the bricolage and the mired relations involved in a space analogue mission in Scotland/ Mars. These creations can bring things into relation in new ways that evoke another place, be it Mars, or the space between Mars and Earth, between imagination and reality; just as the astronauts vacillate between Earth and Mars, the artworks evoke a betweenity.

During the mission, Sarah was able to excavate one complete deer skeleton and several shed antlers (often whilst resident deer looked on) and subsequently fabricated a deer spacesuit (see Figure 9.1). Sarah suggested the spacesuit could be an offering left for the deer who desires protection from alien (i.e., human) encounters. The objects made and the production process itself can focus 'unconscious reasoning' (cf. Marcus Coates' 2013 'The School of the Imagination' project). That is, the process of making, circulating, and engaging with the artworks brings to the fore a sensibility of the place and relations involved in being on Earth/*Mars*. The artist less produces reflections and interpretations of an event but rather brings attention to the new and novel coordinates of existence that emerge through performative acts. Sarah was not simply recording the event of being on *Mars*/Scotland, but actively producing methods through which to think, report, and analyse this act of imagining care in remote environments.

Our third register of bricoleur is the researcher anthropologist. In this volume, the editors have asked how thinking about outer space has troubled our notion of the field site (cf. Johnson; Ojani; Dovey and Potts (Chapters 7, 3, and 8, respectively, this volume; Szolucha et al. 2023). One thread of thought from the contributions to this volume, and the field of the social studies of outer space at large, is that the field site is multi-temporal, multi-scalar, and multi-sited (Gorbanenko, Jeevendrampillai, and Kozel, Chapter 1, this volume). As conceived here, the 'field' is less a place but a phenomenon around which the social and material relations that constitute the process of making a meaningful life are worked out. If we are to study how people construct worlds of meaning and regimes of values, and orientate their efforts through the employment of different registers of time (such as the future, in which people live on Mars), scales (such as where people think of social relations at the scale of the planetary or cosmic),

Figure 9.1 'Stag,' by Sarah Fortais, 2016–ongoing. Materials: French military parachute bag, deer skull, coat hanger, piano key mechanism, Skull Hooker, plastic bags, newspaper, curtains, golf bags, shin guard, elbow pads, ski bag, ducting, boots, running shoes, children's booties, bike tire, reflective bands, leather jackets, horse ankle/shinguards, bedpost, CD rack, plastic tubes, rock wool, film splicing tape, packing tape, grommets/eyelets/metal fixings, nylon/leather/cotton/polyester-nylon rope, parachute cord, wood trolley, Gorilla Glue.

or through multiple places simultaneously (such as through Scotland and Mars), then we also need to think about how our understanding of social relations is produced and circulated in economies of knowledge. We argue that engagements between artists and anthropologists can lend us new and innovative methods in anthropology. Roger Sansi (2020, 5), in his discussion on the relationship between the anthropologist and the curator, has asked what it is that constitutes the 'work' of ethnography. He flags the work of Paul Rabinow and colleagues (2008) who propose that the term 'assemblage work' can replace the term 'fieldwork.' As Sansi states:

> The object of study of anthropology is no longer a given singular community, located in a singular space for a particular time, but an assemblage of different parts: people, places, objects, concepts and agencies of different sorts that constitute contemporary assemblages. This notion of assemblage-work is related to Marcus's para-site (2000) – participatory spaces where multiple divergent agents and agencies discursively interact across geographic, temporal and disciplinary boundaries. The model for the para-site would be the design lab, as an open process that does not just represent the world as it is, but produces and tests new experimental objects. The anthropologist as curator would have the role of mediator in the production of these experimental objects.
>
> (Sansi 2020, 5)

What we present here is a paraethnography, one that emerges through bricolage or assemblage work. Taking an emic approach, we think through the practice of being on *Mars* on a Scottish Island, of being para-sited. In the next section of this chapter, we present the fieldwork before concluding with thoughts on the process and what it might lend us in terms of the development of new modes of anthropological methods.

What happened

It took a day to get to the *launch site* off the coast of mainland Scotland. After sampling the local whisky the night before, we awoke relatively fresh and ready to spend five days on a remote island. A taxi ride took us to a small port from where we had scheduled a chartered boat to take the whole crew and equipment over to the island. The driver made small talk as we city slickers scrambled to find the cash that he asked to be paid in (no cards here, no signal). 'Where are you folks going?' He asked as we avoided the local car ferry to the bemusement of the ferry assistants. '*Mars*' we responded whilst heaving bags to the port side. He took it as it was, 'righto, £21 please.' In two shifts, we travelled to the island, which the day previous a local shopkeeper had described to us as 'untidy.' First, the ten mission crew: a mission lead, mountain leaders, anthropologists, a general medical practitioner (medical supervisor for the whole team), case study actors, and Sarah, our artist in residence. They set up camp, and prepared the simulations. The analogue astronauts, the photographer, and other researchers would arrive the following day.

The crew, who were carefully selected for the mission, had completed their pre-launch training and had been at the launch centre for some days. They had carefully packed their few permitted personal items. They decontaminated themselves of Earth and carried out last-minute preparations for Mars. They said their goodbyes to loved ones over their radios and completed social media sound-bites for the media team. The launch was t-minus 1 hour, they awaited the blast-off. Whilst nervous, they were confident in the ground team's preparations. The shuttle seemed big and powerful, there was a hive of activity occurring around them. The shuttle was loaded with gear. The crew sealed their space suits. The weather for launch was good but turbulence was expected.

A crew of two men, one of whom co-owned the island, arrived with the boat. It was called the Razorbill, also the name of one of the native birds on the island. We donned waterproofs and formed a chain to load the boat with the bags of personal belongings, food, and research kit, which included a lifelike silicone man to be used in one of the role-play medical scenarios. The team sat evenly on either side for twenty minutes through the choppy water and rocky outcrops. The weather was good: warm and overcast.

On arrival, we formed a chain once more to move the equipment over the rugged terrain to the bothy, a small house on the island which served as a base camp for the support crew. The island has been uninhabited by humans for over seventy years. There was no mains water, electricity, or any such creature comforts. There were several herds of deer which kept the vegetation down and provided ample food for deer ticks, which may carry Lyme disease (a bacterial infection caused by tick bites that can result in serious ill health). The mission director gathered the team, gave a safety briefing and a tour of the bothy, and instructed us on using the long-drop toilet facilities (toilet downstream, water collection upstream). The crew tents were set up, the artist set up a workshop inside the house, and a personal hygiene area was established for checking for ticks. The sponsored expedition foods were divided up as per people's pre-mission choices, and we ate our first 1000kcal meal rehydrated and out of a bag.

The following day, the team carried out reconnaissance covering most of the island. It rained a lot. The support team made the final decision on where to best place the base camp for the analogue astronauts. It needed to be sheltered(ish) and far from the mission crew support base, yet easily viewable from a distance. We set up the polar expedition tent that had been loaned for the mission by one of the guides who had experience in the Antarctic. Some aspects of the landscape provided a problem for mission fidelity – for example, an old cabin on the beach interrupted a landscape otherwise free from obvious human activity. After some discussion, we decided that such things could be repurposed and explained as *'cargo from medical drops'* and that we could incorporate the collection of medical supplies from these cabins into mission simulations.

Figure 9.2 'Definitive Proof We're on Mars,' by Sarah Fortais, 2022.

The support crew carried out their rain-soaked tour of the entire island to assess the possible mission simulation sites. Whilst excavating bones in one of the caves, one of the crew unearthed a Mars Bar wrapper buried in the peat. He lifted it aloft and declared, with a sudden emotional fervour, that we *'finally have definitive proof we are on Mars.' The archaeological find was sealed in a makeshift hermetic chamber by Sarah on our return to Earth (see Figure 9.2).*

This artwork, along with subsequent sculptures crafted from the excavated animal bones and flotsam gathered from the island, generated conversation about the mission and reflections on the process of running an analogue Martian simulation. The found objects produced a cohesive visual identity for a mission that is simultaneously historical, contemporary, and future-focused. The resulting sculptures served as a useful device to structure talks about the mission to different audiences (see Figures 9.1, 9.2, and 9.3).

For the astronauts, the journey to Mars was months in the making. Selection, medical, wilderness, and geological training had all led to this moment. They arrived at their selected base camp, affectionately called 'the island,' by the Razorbill rocket, through a turbulent atmosphere.

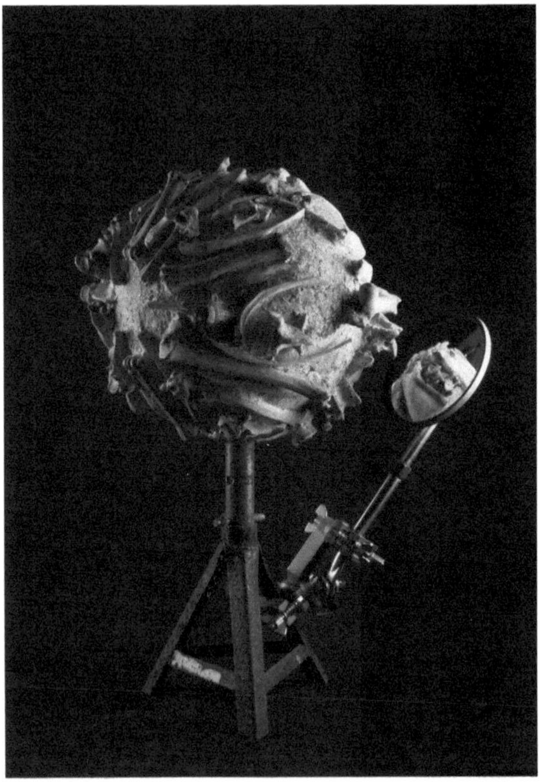

Figure 9.3 'Lunga Ball 002 (Vanity Sheep),' Materials: sheep skeleton, Lunga soil, jesmonite, gorilla glue, car jack prop, drum clamp, scooter mirror by Sarah Fortais, 2022.

After adorning their mission patches, unfurling their maps, and having a safety brief, which included a medical reading of their heart rates by mission control, the astronauts set off across the terraformed Mars towards their base camp past a sign that read 'The island is privately owned […]. Do not leave litter please or start fires or disturb the peace of the island.' Mars was to be studied geologically for future mining missions. Previous crews had already been here, and signs of their trip were visible in the abandoned base camps. These crews had lost contact with mission command some time ago and were assumed dead. The mission briefing of the new crew was to identify (1) potential landing sites for future missions and (2) geologically interesting areas. Their mission began when they received a communication check the following morning from the mission command who were separated by a 40-minute communication delay.

Back at base camp, the team finalised many logistical decisions. Pete, the mountain leader who would keep his eye on the astronauts, held a phone that the astronauts could message using WhatsApp. Pete would then wait 20 minutes before forwarding that message to a phone at mission control, thereby simulating the communication delay (20 minutes each way) likely on a Mars mission. The mission crew and actors finalised the details of how the simulated healthcare emergencies should run. The actors ran an impromptu character development workshop to give themselves a better understanding of the medical details of each scenario and how their characters might respond to crew engagement. The backstories of the previous Martian crew who had been stranded near the base after a failed mission were developed by the actors to increase fidelity and the feeling of being on Mars.

After learning of the island's historical ghost stories from the island's owners, the mission crew spent the night recounting the stories. These stories, just like the mission simulations with the Martian astronauts, involved communication delays, but via ghostly lights from other islands rather than through simulated delays via WhatsApp. The visible mainland coast from the island is the territory of the Scottish MacDougall clan, whose ancestral home at Castle Dunollie contains a seven-sided cupola grown from Ash trees. The afternoon before travel, the structure was re-imagined to mirror the cupola on board the International Space Station as a playful gesture towards the evening's journey to Mars. The crew were using storytelling and creative practice with landscape and atmosphere. Anthropologists have long argued that myth and storytelling tie people to place, kinship, and ethical worlds (Houseman 1998; Basso 1996; Lévi-Strauss 2011 [1955]). Here stories from Scottish history were being reconstituted to create a Martian identity. In this case, storytelling is about preparing and inhabiting a world to come, about getting space ready.

Day one 8:45

*MD: Good morning analogue astronauts. This is mission crew and I hope you slept well. Your first objective for today is to collect the medical equipment that has been sent ahead of you. It was transported in two small modules that have landed North of your location at *Grid Reference 703 085*.*

*The modules had a bit of a bumpy landing so we are unsure what equipment has survived. After the research studies, please go straight to the modules and audit the equipment. *Please send us the list of equipment that has reached you. Once you have achieved the medical equipment audit objective, please continue with the mission aims. Remember to conserve your communication device's batteries – you may need them if there is an emergency* – Mission director out.*

The first morning of the mission proper, the research team hiked to a beach where the mission director had told the astronauts *a medical resupply case had been dropped*. Whilst observing, the mission crew did not speak to the astronauts, avoided eye contact and smiles, and did not give any sense that we were co-present. The only mode of communication with 'mission control' for the astronauts was using WhatsApp. One of the actors lay in the bracken and screamed for help across the bay.

Almost immediately upon commencing their mission, the crew received a distress call from another crew whose ship had broken up in the atmosphere on approach, scattering distressed astronauts across the surrounding environment. A medical kit was airdropped onto a nearby plateau. The astronauts moved towards the signal. Upon encountering a crew member from a previous mission who clearly needed medical help, the medic quickly assessed him and placed him on a stretcher. The astronauts decided to move the patient back to their base camp and had to navigate the difficult terrain whilst maintaining a duty of care to the patient. The downed astronaut was a young man with a broken hip. The journey back to base camp was laborious and it required all five astronauts to carry the patient between them on a stretcher for three-and-a-half hours. The astronauts moved slowly through the landscape. Whilst waiting for the astronauts to navigate a steep route and a small boggy ravine, Sarah was able to excavate a deer skeleton in the nearby bog.

When back at the base, the tick-checking ritual proved extremely important to the astronauts for team-building and developing rapport. The tick checks were perhaps their most effective act of teamwork because they were successfully able to define roles, stick with them, and continuously stay on task. The tick checks aided the transformation of the tent into their 'mission base.' *The base had to be decontaminated regularly.*

On the second night at base camp, the astronauts received a distress call. After venturing out to follow the beacon's signal over the Martian terrain, they found the commander of another crew nearby the basecamp. They had sustained serious trauma wounds and a splenic rupture which was causing serious internal bleeding. The patient's condition was slowly deteriorating. The astronauts also found a second patient nearby. Whilst they were psychologically distressed and had difficulty breathing (with a tension pneumothorax), they were more concerned for their commander's well-being. Throughout the night, patient 'A' slowly passed away. The astronauts had to make vital decisions on how to provide care to a patient who had little chance of surviving. On Mars, how to deliver care, use limited resources, and deal with a dead body is radically different to how it is in most Earth-based contexts.

Ahead of the mission proper, the mission leader visited the island to assess the suitability of the site. Sarah had requested that he retrieve soil samples so that she could test their properties. The highly acidic and peaty soils turned out to be a good material to mix with Jesmonite, a material often used in lifecasting. After some experiments, the artist was able to make a casting material that was up to 95 percent soil from the site. This was very important as the castings were to be made on the island. This use of in situ resources was conducted in the spirit of preparing for life on a remote planet where non-local materials were either impossible or hugely expensive to get. The artist was also considering the forms of ritual practice that might be needed on Mars. How to deal with a dead body may be among them.

There were multiple reasons to cast a death mask. The 40-minute process of casting served as a ritual to collectively mourn in a remote, resource-scarce environment. In an off-world environment, a human body would likely be too environmentally valuable or expensive to send back to Earth. The process of creating a death mask (or hand cast) (see Figure 9.4) could bridge cultures and promote a feeling of connection through familiar methods and artistic output. Sarah updated the practice with new material recipes and re-purposed the waste mould-making materials in fertiliser and other artworks (see Figures 9.3 and 9.5). Sarah's innovations made this practice more sustainable and thus appropriate for both off-Earth and climate emergency conditions. Using soil and biological material extracted from the casualty site reduced the amount of material to initially pack and enabled Sarah to produce a range of colours unique to that environment. *The resulting site-specific casts are physical and emotional records that could be sent back to Earth.*

The morning after the patient died, the astronauts were sleep-deprived and physically tired. They were given land-surveying exercises to complete by mission control. At 9am Martian time they were informed that due to an impending solar storm, they would need to abandon their camp with all their belongings, and within eight hours, they would need to arrive at the launch area ready for their return journey. However, on their journey the astronauts would be bombarded by radiation from the solar storm for at least 10 minutes every hour; they would need to be careful.

They were given a radiation shield (an orange portable rain tarp). A radiation warning was given from mission control. The mission leader was hoping the astronauts might engage with the landscape in creative ways, using the local caves along the coast as shields. *The astronauts were slow in their movements, but eventually, they came across another distressed crew member from a previous mission. They assessed the patient as having a sprained ankle and began to make their way, with the patient, to the launch site. The final 500 metres proved difficult for the overtired astronauts.*

Figure 9.4 'Martian Death Mask,' by Sarah Fortais, 2022. From the Remote Care Case series of antique first aid kits repurposed to contain life castings taken while on the Lunga 6 Analogue Space Research Mission. Materials: First Aid kit, aluminium dibond print, life castings made from Lunga soil and Jesmonite, Lunga 6 patch cast from Lunga soil and jesmonite, found objects.

The walk to the bothy *(launch site)* was a relatively straight slow decline through open wet fields. However, fatigue resulted in hasty decision making in navigation. They had carried their gear at the start of the analogue mission with little effort, but now at the end, with the added weight of the medical pack along with the needs of a hobbling patient with

Figure 9.5 'Lunga Ball 016 (Soft Sculpture),' Fortais 2022. The artwork is displayed with the following text from the mission transmissions: *08:39: WARNING – we have had notification of a solar storm heading your way. Your base camp is in a danger zone and not suitable for life for prolonged periods. Every hour on the hour from 0900 your local time there will be a solar flare with risk-to-life radiation. You must seek shelter for five minutes, after which you can go outside until the next hour.*

a sprained ankle, they found they could not manage the weight. At about 4pm, an hour before the end of the mission, the team decided to split up when *the final injured crew member was discovered close to mission control*. By this point, crew dynamics were strained; people were tired, cold, and wet, and the mission was nearly over. In moments of pause from carrying bags, the astronauts would slip in their language from being in character on the mission, talking about Martian landmarks and mission aims, to talking about getting home, and how they were looking forward to their shower and bed. This 'Earth talk' allowed the crew to maintain relations whilst mission dynamics were strained. Vacillating between *Mars* and Earth allowed them to navigate social tensions and demonstrate that tensions were not personal but were to be left on *Mars*.

Conclusion

This chapter has presented a paraethnography of simulating healthcare protocols in remote and extreme environments by being on Mars whilst on a remote Scottish island. The ethnographic style emerges from the field that vacillates between places and through being in landscapes multiple. We foreground the method of bricolage in our analytic frame, ethnographic method, and our mode of engaging with the act of making the island Martian. For us, bricolage is not about building towards an a priori imagined finished thing, but rather creatively fashioning joints thereby embedding oneself into the process of making art and anthropological knowledge. As we re-imagined old huts as medical drops, ticks as radiation, and chocolate bar wrappers as Martian archaeology, we brought attention to connections, happenings, and moments. This served to build a collective sense of being in a particular place, Earth, *Mars*, or somewhere in between.

Recent discussions, building on the debates of the reflexive turn in anthropology of the 1980s and '90s (cf. Clifford and Marcus 1986), have argued that we must go beyond representing a distant 'other' (Schacter 2020; Jeevendrampillai 2021), be open to new forms of interruption and disruption (O'Neill 2016; Flynn 2020, in Sansi 2020), and engage 'in experimental, speculative, long-term processes wherein we can speak together with our interlocutors, mediating not controlling their own ways of seeing' (Schacter 2020, 202). Here, not only are we mediating a representation of an a priori world, but rather we recognise that we are also co-producing worlds in the very acts of doing anthropology. The anthropologist here is much like the artist: we both reflect upon, draw attention to, and make relations. We world-build. There is an openness to bricolage as a method. As Les Roberts (2018, 13) asserts, bricolage 'can help make explicit the poetics and affects of space, as well as the ethical and procedural frameworks that are brought to bear on how space is put into practice.' Joe Kincheloe has noted that bricolage as a method has often been seen as 'superficial' (Kincheloe 2001, 680–681) due to its openness. Yet, he argues, it is this openness that lends bricolage its rigour. By virtue of the unique connections, happenings, and moments, the methodology produces continually novel results. One might use one's sensibility to gravitate toward the same things (animal bones or travel talk), but the environment can evoke, surprise, and detour one's thoughts and actions.

Responding to the issues outlined in the introduction, this chapter asks what sort of methodologies are needed to deal with a field conceived less as a place than as a phenomenon that is multiple in time, scale, and location. Our methods of analysis and knowledge exchange, such as the production and display of artwork (to take a wider view than simply writing), require rethinking and decoupling from the notions of the real, boundedness, and linearity in respect of the fieldsite (cf. Martínez 2021). In moving from 'critical thinking' to what Salazar and Gorman call 'critical making' (2023, 2) we advocate for experimental collaborations (cf. Jeevendrampillai 2017). As Tomás Sánchez Criado and Adolfo Estalella (2018, 1) note, experimental collaborations compel us to 'reconsider not only the epistemic practices, types of relationships and forms of engagement in our fieldwork, but also our accounts of the field.' Such an approach 'de-centers the conventional ethnographer-informant relation through a para-ethnographic epistemic partnership' (Boyer and Marcus 2020, 13). But further to this, decentring might reflect the process by which knowledge is produced through the idea of the field. As astronauts collected information, the crew dug bones, stories were told, and as tea leaves and gauze were turned into sculptures, Mars as a place and Tartan Martians as a vacillatory subjectivity were brought into being.

Acknowledgements

This chapter was produced with the explicit help of Myles Harris, Aaron Parkhurst, and Adryon Kozel, who all were involved in both the mission itself and in the early brainstorming sessions as to what this paper might look like. Many of the ideas in this paper, namely the idea of writing multiple, come from those collective discussions. This publication is part of the ETHNO-ISS project that has received funding from the European Research Council (ERC) under the Horizon 2020 research and innovation programme (Grant agreement No: 833135). The chapter was also funded by UCL's Centre of Outer Space Studies, and the UCL Grand Challenges and the co-operation and access granted by Space Health Research. Thank you to Jennifer Christy for giving comments on the draft.

References

Austin, John Langshaw. 1975. *How to do things with words*. Harvard University Press.

Basso, Keith H. 1996. *Wisdom sits in places: Landscape and language among the Western Apache*. UNM Press.

Boyer, Dominic, and George E. Marcus. 2020. "Introduction." In *Collaborative anthropology today: A collection of exceptions*, edited by Dominic Boyer and George E. Marcus, 1–21. Ithaca, NY: Cornell University Press.

Clifford, James and George E. Marcus (eds). 1986. *Writing culture: The poetics and politics of ethnography*. Berkeley: University of California Press.

Coates, Marcus. 'School of imagination' 2013. www.marcuscoates.co.uk/projects/95-school-of-the-imagination. Accessed 8 April 2024.

Derrida, Jacques. 1988 [1971] *Signature, event, context*, trans. Samuel Weber and Jeffrey Mehlman, in Limited, Inc. Northwestern University Press.

Estalella, Adolfo, and Tomás Sánchez Criado (eds). 2018. *Experimental collaborations: Ethnography through fieldwork devices*. Vol. 34. Berghahn Books.

Flynn, Alex. 2020. "The curator, the anthropologist:'presentialism' and open-ended enquiry in process." In *The anthropologist as curator*, pp. 173–193. Routledge.

Fortais, Sarah. 2018. “Defining cool through a Bricoleur’s studio practice.” UCL Doctoral Thesis 2018. https://discovery.ucl.ac.uk/id/eprint/10052124/
Hage, Ghassan. 2009. “Hating Israel in the field: On ethnography and political emotions.” *Anthropological Theory* 9(1): 59–79.
Hage, Ghassan. 2017. “Diaspora and migration.” In *A companion to critical and cultural theory*. Edited by I. Szeman, S. Blacker, and J. Sully. John Wiley and Sons, pp. 191–204.
Houseman, Michael. 1998. “Painful places: Ritual encounters with one’s homelands.” *Journal of the Royal Anthropological Institute* 4(3): 447–67.
Jeevendrampillai, David. 2017. “Playing with encounter.” *Suomen Antropologi: Journal of the Finnish Anthropological Society* 42(3): 55–57.
Jeevendrampillai, David. 2021. *Citizenship, Democracy and Belonging in Suburban Britain: Making the Local*. UCL Press.
Jencks, Charles, and Nathan Silver. 2013. *Adhocism, expanded and updated edition: The Case for Improvisation*. MIT Press.
Kincheloe, Joe L. 2001. “Describing the bricolage: Conceptualizing a new rigor in qualitative research.” Qualitative Inquiry 7(6): 679–92.
Lambert-Beatty, Carrie. 2009. “Make-believe: Parafiction and plausibility.” *October* 129: 51–84.
Lévi-Strauss, Claude. 2011 [1955]. *Tristes tropiques*. Penguin UK.
Lévi-Strauss, Claude. 1962. *La Pensée savage*. (The Savage Mind). Translated from the French by Anonymous. George Weidenfield and Nicholson Ltd.
Marcus, George E. (ed.). 2000. *Para-sites: A casebook against cynical reason*. Vol. 7. University of Chicago Press.
Martínez, Francisco, Eeva Berglund, Rachel Harkness, David Jeevendrampillai, and Marjorie Murray. 2021. “Far away, so close: A collective ethnography around remoteness.” *Entanglements* 1(4): 246–283.
Mol, Annemarie. 2002. *The body multiple: Ontology in medical practice*. Duke University Press.
O’Neill, Paul. 2016. *The culture of curating and the curating of culture(s)*. MIT Press.
Rabinow, Paul, and George E. Marcus. with James D. Faubion, and Tobias Rees. 2008. *Designs for an Anthropology of the Contemporary*. Duke University Press.
Roberts, L. 2008. “Spatial bricolage: The art of poetically making do.” *Humanities* 7(2): 43.
Rose, Gillian, Monica Degen, and Begum Basdas. 2010. “More on ‘Big Things’: Building events and feelings.” *Transactions of the Institute of British Geographers* 35(3): 334–49.
Salazar, Juan Francisco, and Alice Gorman. 2023. “Social studies of outer space: Pluriversal articulations.” In *The Routledge handbook of social studies of outer space*. Edited by idem. Taylor & Francis.
Sansi, Roger (ed.). 2020. *The anthropologist as curator*. Bloomsbury Academic.
Schacter, Rafael. 2020. “A curatorial methodology for anthropology.” In *Lineages and advancements in material culture studies*, pp. 190–204. Routledge.

Szolucha, Anna, Peter Timko, Chakad Ojani, and Karlijn Korpershoek. 2023. "Ethnographic research of outer space: Challenges and opportunities." *Ethnography* https://doi.org/10.1177/14661381231220273

Tilley, Christopher. 1994. *A phenomenology of landscape: places, paths, and monuments*. Vol. 10. Berg.

Tilley, Christopher, and Kate Cameron-Daum. 2017. *An anthropology of landscape: The extraordinary in the ordinary*. UCL Press.

10 Composing the cosmos

Tuning into multiplicities with Thai Buddhist concepts

Lauren Reid

Inside the Milky Way

"How can we know what is on the other side of the epicentre of the Milky Way? What is its structure?" These questions drive Supachai, an astronomer working at the National Astronomical Research Institute of Thailand (NARIT), who researches exoplanets and the structure of the Milky Way. I met him during fieldwork in Chiang Mai. With a constant smile and slight lisp, he explained to me how new technologies have radically opened up possibilities for mapping the vast spiral galaxy in which our solar system is situated. However, due to Earth's position in one of the galaxy's multiple spiral arms (known as the Orion Arm), within an immense milieu of stars, gas, dust, and dark matter spanning a diameter of about 100,000 to 120,000 light-years, it is incredibly difficult to determine the structure, mass, and radius of the Milky Way's central bulk. Even more challenging is the prospect of finding out what it looks like from the outside, above, or below. As Supachai succinctly puts it, "We cannot look into the other side of the Milky Way because we are *in* the Milky Way."

The challenges posed by studying and imaging the Milky Way "from within" raise key questions: how do situated positions and conditions affect the kinds of information produced? How might one overcome the limitations of current knowledges, methods, and technologies when studying vastly different worlds? How to comprehend multiplicities of scale and time that go far beyond familiar experience? Are there ways of getting "outside" the very thing that we are in the middle of?

These questions echo central concerns in the field of anthropology, particularly in regard to the challenges that anthropologists face in studying different lifeworlds from an external vantage point. It has long been recognised that "ethnographic knowledge is always situated, positioned and constructed," shaped by the researcher's accompanying presuppositions (Stodulka 2023, 105; see also Clifford and Marcus 1986; Haraway 1988). Anthropologists and social scientists more broadly underscore the

DOI: 10.4324/9781003437956-10

necessity of taking the conditions that shape their research into account by, for example, incorporating reflexivity to cast light on the researchers' own stakes and epistemological limitations (Verran 2014), decentring or provincialising dominant perspectives (Chakrabarty 2000; Spivak 1988), and advocating for understandings of pluriversality and multiple realities to counter assumptions about the uniformity of conditions across different contexts (Escobar 2020; Law 2015).

A significant point of anthropological contention is how foundational concepts like "culture" (Descola 2013; Wagner 2016), "nature" (Kohn 2013; Viveiros de Castro 2014), "belief" (Aulino 2022; Mair 2012), and many more supposedly universal terms, turn out to be historic inventions that arise from anthropology's European origins. The uncritical persistence of such analytic categories, in turn, reproduces the biases and presuppositions of the Euro-Christian roots from which they primarily grew (Latour 1993 [1991]; Rees 2018). Marilyn Strathern famously argues that concepts do not simply point to external referents but are active entities that create worlds of their own (1988). To take an example from cosmological history, when the Earth was regarded as the centre of the universe, the cosmos was composed in such a way that all other entities and phenomena seemed to orbit it. Yet when heliocentric models emerged, the cosmological order was dramatically rearranged so that new formations and relations took shape. As Marisol de la Cadena succinctly puts it, "it matters what concepts we use to think other concepts" (2015, 27).

Human engagements with outer space amplify the complexities of working across multiple scales, reckoning with alterity, and imagining otherwise. As Milky Way researcher Supachai knows well, common concepts and Earthly conditions cannot be projected onto the unfathomably complex Milky Way because the material conditions of the cosmos (air composition, water content, magnetic fields, planetary orbit and rotation, proximity to the sun, etc.) are entirely different from Earthbound ones. Cosmic engagements pose the challenge to operate and think radically otherwise, for space scientists and anthropologists alike (Battaglia 2012; Buchli 2021; Mirmalek 2020; Valentine 2017). In this sense, thinking with outer space in anthropology (or extending anthropology beyond Earth) raises the stakes for addressing set modes of conceptualising and producing knowledge about different worlds.

This chapter addresses the quandary of how taken-for-granted concepts and situated assumptions shape the cosmologies in which we operate. I turn attention to a script about an imagined future encounter in outer space that was created during my fieldwork by a Thai ufological Buddhist and a Dutch curator. While the pair jointly wrote a cohesive script, it can be read in different ways, depending on one's underlying cosmological framework. I will explore understandings of key concepts within the script of "selfhood," "nature," and "existence," based on insights gained through time spent with a group of Buddhist extraterrestrial-contactees and drawn from academic scholarship on Thailand and Buddhism. By integrating Thai-Buddhist-derived concepts into the script's narrative, I will show how local specifics within space-focused communities in Thailand offer one among many ways of composing and relating to the cosmos. Here, I make the case that engaging with multiplicity by reconfiguring various elements such as concepts offers a generative approach to apprehending other worlds and possibilities.

A view from Thailand: Multiplicities and hybridity

I first want to describe how and why the script I will analyse came to be. It was created towards the end of my fieldwork in 2019. I had spent the previous twelve months at different locations in Thailand, interviewing researchers in space science institutions (including the aforementioned Milky Way researcher, Supachai); participating in activities of an extraterrestrial-contactee group; and observing space science outreach centres. The aims were twofold: to understand how human space exploration (in the broadest sense of actively engaging with outer space) is practised and imagined, and to consider how emergent space-focused communities may open up alternative understandings of human relations off Earth.

Thailand may seem like an unusual location for studying outer spatial practices since it is not at the forefront of space exploration. Present-day space endeavours are dominated by national space science agencies that increasingly include non-Euroamerican projects from China, India, United Arab Emirates, among others, as well as billionaire-led commercial ventures such as SpaceX, Blue Origin, and Virgin Galactic. However, they represent only a fraction of the complex network of contemporary relations to outer space that unfold across a variety of contexts such as telescopes in Hawaii (Casumbal-Salazar 2017; Swanner 2017), satellites orbiting the Kármán line (Barker 2005; Gärdebo et al. 2017), and space artefacts in exhibitions (Dovey and Potts, Chapter 8; Osbourne, Chapter 5; Tereshin and Sivkov, Chapter 6, all this volume). I consider Thailand an instructive case because the space science sector only recently gained momentum in the early 2000s, following the establishment of its first publicly funded astronomical institute, NARIT, in 2009. This fledgling yet rapidly developing field of space science presents a de-centred context from which to observe the flows of space-scientific practices and knowledges in relation to various local ways of knowing and being in the cosmos.

In tandem with this emergent space science sector, a proliferation of Buddhist and animist ontologies throughout Thailand offers a pluriversal array of cosmic orders to tune into. After all, there is more than one way to conceptualise outer space, as scholarship on sky knowledge shows (Geertz 1984; Holbrook et al. 2008; Ruggles et al. 1993). In this volume, Anne W. Johnson (Chapter 7) emphasises wide-ranging understandings of the Moon by drawing insights from a "space milieu" in Mexico that spans artists, NewSpace entrepreneurs, museum-goers, astrophysicists, and others. Furthermore, Hae-Seo Kim (Chapter 11, this volume) shows how various cosmologies interrelate by highlighting the interplay between scientific, shamanistic, and astrological cosmologies in shaping South Korea's "Space Age." These cosmic multiplicities are profoundly embedded in fundamental presuppositions about how our realities work.

Despite a multiplicity of cosmologies, human engagement with outer space is typically regarded as a technoscientific problem (Traphagan 2020), while science is posited as a largely objective and detached enterprise by persistent modern philosophical frameworks. Social studies scholarship, however, has long demonstrated how the sciences are embedded within social, ontological, and historical contexts (Knorr-Cetina 1981; Harding 2008; Kuhn 1962; Latour and Woolgar 1986) that are animated by intricate relations between infrastructures (Larkin 2013; Ojani, Chapter 3, this volume), imaginaries (Jasanoff and Kim 2015), performative practices (Bunch, Chapter 2, this volume; Pickering 2017), and myriad other material and immaterial phenomena and conditions. Following such heterogeneous and co-productive interrelations, I share the notion argued by such scholars that the way we live is "always and already hybrid" (Bauman 2015, 389).

For these reasons, this research was multi-sited, spanning communities from what could broadly be described as astronomy, ufology, and Buddhism. Here I share Peter Timko and Karlijn Korpershoek's assertion in this volume (Chapter 12) that "grappling with the shifting multiplicities of space calls for making nimble methodological changes to our research activities, particularly in how we construct space as an object of study in our fields" (see also Jeevendrampillai and Fortais, Chapter 9, this volume). I consider a constant shift in overlapping perspectives to be vital to composing a picture of outer spatial practices and future imaginaries in Thailand without falling into simple essences.

The group UFO Kaokala, a central community of this research, epitomises the "shifting multiplicities" and "always and already hybrid" spirit. Its followers believe in extraterrestrial intelligence and attempt to commune with extraterrestrials via Theravada Buddhist concepts and practices such as meditation. Group members strive to improve their Buddhist practice so that they can release their affective attachments, break cycles of rebirth, and, ultimately, transcend the planetary constraints of Earth. At the same time, figures and aesthetics from Hollywood blockbusters, such as *The Avengers* and *X-Men*, are oft-repeated references. The group also frequently uses language and concepts from quantum physics and understands the human mind as though a "technology" that exists within a larger "matrix" or "system" of consciousness. This networked mind can be enhanced or "upgraded" through cognitive training, an approach that would not be out of place among the Transhumanist and Singularitarian philosophies popular in Silicon Valley. UFO Kaokala's creative and hybridised mixing of components (extraterrestrial intelligence, Buddhism, technoscience, pop culture) destabilises taken-for-granted monolithic conceptual categories such as "science," "religion," "Western," "Thai," and so on.

As I continued my fieldwork across diverse space-related sites in Thailand—meditating with UFO Kaokala members atop a mountain known as a "stargate" for extraterrestrial communication, visiting NARIT's cutting-edge telescope and spectrograph used to find exoplanets—the focus on multiplicities and hybridity prompted me to imagine potential conversations among the various informants I encountered. I wondered whether it would be possible to bring individuals from these distinct sites into direct dialogue. I hoped to generate "ethnographic situations" in a style akin to what Nikolai Ssorin-Chaikov describes as "ethnographic conceptualism." Ssorin-Chaikov proposes that "In contrast to ethnography as participant observation of what exists, ethnographic conceptualism explicitly constructs the reality that it studies" (2013, 8). Following this thinking, rather than obscuring the situated and co-productive nature of anthropological research, creating ethnographic situations makes it overt.

Additionally, since the topic is related to human futures in outer space, I wondered what kinds of imaginative scenarios individuals might envision. While throughout my field research, I had largely relied on anthropology's prime methods of direct observation, deep immersion, and interviews, the more speculative aspects of the project called for a multi-modal methodology that could be co-creative (Gubrium and Harper 2013; Smith et al. 2016), imaginative (Dovey and Potts, Chapter 8, this volume; Sjöberg and D'Onofrio 2020), and multi-sensory (Pink 2015).

In response to these two lines of thought, I developed and ran a workshop in which participants were invited to work together to create a short film about a fictional encounter in outer space. With the support of an open call via the host of the workshop—Subhashok Art Centre in Bangkok—and through my own invitations, 18 people took part. Participants included UFO Kaokala members, an astronomer, and a young Buddhist monk, along with curators and artists. Since the workshop combined both local and international attendees, it was run in English with the support of a Thai translator. Over a half-day, the group discussed creative and ethical questions related to space exploration, spanning the playful ("What might the Moon smell like?") and the philosophical ("How might religions be affected by new discoveries in outer space?"). The group discussion was then followed by the creation of a 5-minute-long film about an extraterrestrial encounter. Given film's ability to create impossible leaps through time and space, and to depict images, sounds, and sensations that are far removed from everyday reality (see Lempert 2014; Russell 1999; Suhr and Willerslev 2012), it is a well-suited medium to this project's un-Earthly subject matter. To facilitate closer dialogue between participants, I divided them into smaller groups, each tasked with making one component of a future encounter in space: a soundscape, an atmosphere, a landscape, and a voiceover. In the end, all the elements were brought and performed together in front of a camera to produce a 5-minute film vignette that was a surprise to all.

It is the voiceover script for this short film that is the focus of this chapter's analysis.

A future encounter in outer space

The script was written and narrated by a Dutch curator and a UFO Kaokala group member, Ann. The curator and Ann met for the first (and I believe only) time on this day. This is important because they did not know each other and their accompanying lifeworlds well. The script that they wrote together is a dialogue between two entities, which I present below. As you read it, imagine each voice spoken with heavy echoes and a soundtrack of bleeps and bloops behind them:

A: Welcome, greetings
B: Hi, hello
A: Who are you?
B: I'm the same as you
A: Who am I?
B: A thought
A: Can I think?
B: Why not?
A: Am I your thought?
B: You have your own existence
A: What if I disappear?
B: Don't we all disappear?
A: Are you afraid of disappearing?
B: We don't really disappear, we just change
A: Change into what?
B: Into nature
A: Am I nature?
B: Yes, you are

A: Who made me? Did you?
B: No, you're from another world
A: Are you afraid of me?
B: Why would I be?
A: Should I be afraid of you?
B: It depends, what are you doing here?
A: Exploring exoplanets
B: For what purpose?
A: To colonise for humans
B: Are you good?
A: Only to humans
B: Are they good to you?
A: I don't know, I'm a robot

The story starts with two unknown entities meeting for the first time. Character A tries to figure out who or what they are: "Am I your thought?" "Am I nature?" "Who made me?" The twist comes toward the end, when we learn that A is actually a robot sent by humans to find colonisable exoplanets. Presumably, character B is an extraterrestrial entity in the cosmos "from another world" discovered on A's search for a human-habitable exoplanet. The story then is one of human space exploration, through a nonhuman (robot and alien) perspective.

Before fieldwork in Thailand and as an Australian based in Germany, I would have assumed that the script's storyline follows a kind of Hollywood sci-fi narrative about an extraterrestrial encounter with subtle moralistic undertones of the implications of humans colonising other planets. However, my understanding shifted during time spent with UFO Kaokala members like Ann, revealing tensions between our overlapping realities. Upon closer inspection, concepts within the script can be interpreted in different ways depending on one's ontological assumptions. In the following section, I will work with Theravada Buddhist logics, as understood and developed by the UFO Kaokala members, to elaborate on three themes emerging from the script: selfhood, nature, and existence. This reading provides a conceptual basis for tuning into a multiplicity of ways to know and be in the cosmos.

I refer here to Theravada Buddhism because it is the religious tradition followed by UFO Kaokala and claimed by approximately 93 per cent of Thailand's population. However, it is important to qualify that, like all religions, "Theravada Buddhism" is neither singular, static, nor unified. Thailand's religious systems have been characterised by years of dynamic mixing with different traditions—Hinduism, Brahmanism, Indigenous animism, and Chinese folk religion. How to make sense of the complex and rapidly changing hybrid nature of the religious landscape in Thailand has been the subject of decades of scholarly debate (see Kirsch 1977; Kitiarsa 2005; Terwiel 1976). UFO Kaokala exemplify this complexity through their hybrid worlding of ufology with Buddhism. The following elaborations, therefore, can only be partial and roughly sketched. To broadly trace the contours of each concept, I rely on various teachings published by the group written by Ann and two of its three leaders, Wassana and Somjit, as well as Thai studies scholars.

On selfhood

A (robot): Who are you?
B (ET): I'm the same as you
A (robot): Who am I?
B (ET): A thought

For those familiar with Euro-modern traditions, it is hard not to hear Descartes' famous statement "I think, therefore I am" in the excerpt above. However, rather than a Cartesian body-mind self, UFO Kaokala members follow Theravada Buddhist teachings that humans are composed of five aggregates: physical form (including the body and its sense organs), sensation (or feeling, which can be pleasant, unpleasant, or neutral), perception (or recognition, which involves recognising objects based on past experiences), mental formation (like thoughts, emotions and intentions), and consciousness (or awareness, the foundation for all mental activity and experience).

It is the goal of Theravada Buddhist practitioners to recognise that these aggregates are impermanent and, thus, there is no inherent self—no "I." Or as UFO Kaokala's co-leader, Somjit Raepeth explains, "the five aggregates don't really belong to us [...] All beings must reach the point in which we experience ourselves as non-self" (Raepeth 2014). The (non)self or *annatta* contrasts with Euro-modern logics that consider the self a stable, bounded, and continuous entity (Lambek 2013). Instead, the non-self is constantly changing and "assembles continuously out of conditions" (Stonington 2020, 761; see also Cassaniti 2015; McMahan and Braun 2017). Felicity Aulino, who conducted ethnographic fieldwork with communities in northern Thailand, likens the mind to a turning kaleidoscope in which different "component elements [are drawn together] in an endless array of richly textured combinations. Of course, certain patterns arise with greater frequency and the ever-changing unfolding of existence is in turn shrouded by a powerful illusion of continuity" (Aulino 2022, 232; see also Aulino 2020).

On existence

B (ET): You have your own existence
A (robot): What if I disappear?
B (ET): We don't really disappear, we just change

Connecting the aforementioned concept of "non-self" with "existence." Ann of UFO Kaokala elaborates that "emptiness and [the] non-self do not mean that we really disappear but instead disperse again as particles in the cosmos." Underscoring a sense of component parts momentarily coming together, she says that "even a planet is formed by many particles of dust [...] Earth can be empty in the sense that it needed to emerge from particles" (*The Key*, script written for a short film produced with the author during fieldwork, Thongcharoen, Buranasi, and Reid 2019). In this existential framework, nothing is permanent or fixed and everything is subject to dissolution and transformation so that all things—whether physical or mental—are in a constant state of flux and change.

Such continuous kaleidoscopic transformation is oriented toward cyclical time, in which the universe and human existence undergo continuous cycles of birth, death, and rebirth. Contra to Euro-Christian cosmologies that broadly plot existence along a linear timeline of cumulative progress, Theravada Buddhist cosmologies approach existence as an "unending cycle of growth and decay, integration and disintegration" (Ratanakul 2007, 234). Or in Ann's terms, "all perceptions in any timeline are still just particles and energy that appear and transform into situations" (*The Key*, Thongcharoen, Buranasi, and Reid 2019).

On nature

A (robot): Change into what?
B (ET): Into nature
A (robot): Am I nature?
B (ET): Yes, you are

When existence and the non-self are transient, kaleidoscopic, and interdependent, inevitably, so too is nature. Theravada Buddhist teachings often depict this interconnectedness as a web of dependent origination, where all phenomena arise due to causes and conditions. This nature then is not separate from human beings in the way that it tends to be in Euro-modernist philosophies, where it is associated with wilderness and landscapes, untouched by human development, technology, culture, and progress. UFO Kaokala co-leader Wassana, channelling an extraterrestrial being, wrote that "nature always reflects the minds that live upon it. If the mind is lesser, is not evolved, nature will be variant [...] Those places that need water for agriculture will be dry, but those places that don't need it will be flooded" (Chuensamnaun 2014). Here, the ET-via-Wassana presents a relational metaphysics in which "nature" is integrated with the "minds" of the beings that inhabit it and is shaped by their accumulated karma.

Singular reality troubles

The understandings of "nature," "existence," and "selfhood" are but a few of the script's concepts that I could sketch through UFO Kaokala's extraterrestrial-infused Theravada Buddhism. Additionally, I might have explored other concepts in the script such as "world," "disappearance," "change," and even "robot." It's important to note that each concept carries a variety of associations and inflections that vary world-to-world, person-to-person. Although there is no singular interpretation or common meaning, the intention is to estrange normative Euro-Christian conceptual dualisms like nature-culture, animate-inanimate, self-other, body-mind by presenting one among many possible different metaphysical orientations.

What is at issue here is that two people (Ann and the curator) wrote a singular script via a common language (English), yet at least two rather different conceptual frameworks were potentially being actualised at the same moment. Reading with Euro-Christian historical frameworks, the script can be read as the story of an ET encountering a robot who is searching for both natural wildernesses for humans to tame and a personal sense of individual self. Through Theravada Buddhist frameworks, although the robot is invested in understanding who or what it is, the script is closer to a didactic explanation of cycles of life and rebirth, based on a cosmology of profound interdependence and transience. This matters because although something may seem to belong to a familiar way of knowing and being on the surface, this may not necessarily be so.

The dual worlds of the script pose no significant consequences for Ann and the curator. However, when working with different worlds from situated positions as anthropologists do, and Milky Way researcher Supachai does, unquestioningly projecting one's established presuppositions onto other worlds makes it difficult to recognise, let alone engage with, a multitude of disparate ongoing realities.

I emphasise "realities" in the plural here because a key issue within Euro-modernist logics is the fundamental presupposition that there is a singular totalising reality onto which multiple perspectives are projected—only one of which can be right (Latour 2005; Law 2015; Stengers 2010). Consequently, perceived differences tend to be minimised or disregarded as simply "beliefs" or "representations." However, theorists of cosmo- and pluriversal politics have insisted that there is not necessarily one all-encompassing, objectively measurable reality "out there" (e.g., Blaser 2016; de la Cadena 2015; Omura et al. 2019; Viveiros de Castro 2004). Instead, realities are multiple and enacted through material-semiotic practices (Haraway 1997). Taking seriously the possibility of an abundance of entangled realities, therefore, underscores the particularities of modes of being that claim singularity or universality. As Mark Jackson asserts:

> When we do the work of pluralising the politics of representation, we also see that the ways others always already world their sensibilities are not commensurable with, let alone available to, the imaginaries of dominant Euro-American (and globalising) humanist tropes, or their ontologies, or their consequent politics.
>
> (2018, 10)

If, as such pluriversal scholars contend, realities are composed of complex relations between a wide array of living beings, environments, materials, phenomena, and concepts—such as a bounded self, linear existence, and a separate nature from humans—then those intricate relations, in turn, affect the technologies we develop, the subjects we study, and the issues we grapple with. As Mary Jane Rubenstein writes in her book *Worlds without End: The Many Lives of the Multiverse,* "the shape, number, and character of the cosmos might well depend on the question we ask it" (Rubenstein 2014, 235). Thus, remaining unaware of other possibilities presents significant issues when working with the unknown, as one continues to project and reproduce their same worlds onto difference. More productively, anthropological tuning into and critically analysing a wide array of relations, entities, and environments offers a generative approach to asking new questions and inhabiting worlds otherwise (Bauman 2015).

Conclusion: The Milky Way in motion within the pluriverse

Let us think again with the Milky Way and the astronomer Supachai's research of its structure. From Earthly locales and through human eyes, the celestial bodies in the night sky appear relatively static, even though we theoretically know that the Earth, the planets, and the stars are all in motion. Recent technological developments offer new locations from which to gather unprecedented information about the Milky Way, such as the Square Kilometre Array of telescopes, which (when complete) will span primarily South Africa and Australia, providing an angular resolution higher than any single telescope (see Chinigò and Nieber, Chapter 4, this volume), as well as the Gaia spacecraft that was launched into space in 2013, designed to map nearly two billion objects in the galaxy. Such technologies open expansive possibilities for charting and viewing our galaxy from afar. However, the impression of cumulative progress toward a comprehensive map of the Milky Way is rendered somewhat fantastical if we consider this description of the Milky Way from *Scientific American*:

> We find that the Milky Way is spinning at 236 kilometres per second, which is about eight times the speed at which Earth orbits the sun. Based on these parameter values, we find that the sun circles the Milky Way every 212 million years. To put this in perspective, the last time our solar system was in this part of the Milky Way, dinosaurs roamed the planet.
>
> (Reid and Xing-Wu 2020)

The extract from *Scientific American* recognises that the Milky Way remains incomprehensibly vast and in motion, spinning at great speed in a complex choreography of components, which have their own dynamic systems of interdependent and ever-emerging elements. This rendering of the Milky Way underscores the ways in which our technologies are built for and from a different world that are far from being able to capture even a subsection of this complexity.

I propose, however, that this description of the Milky Way resonates with the Theravada Buddhist worlding that I sketched from the script. Worlds assemble through various components and the relations between them, rather than inhabiting stable and fixed categories based on Euro-modern frameworks (self, nature, robot). This kind of ontology is emergent and generative, since it is based on the existential form of components assembling continuously into new formations and then disassembling at other times.

From here, one can consider how plugging in different components generates novel questions, insights, and even technologies. In a footnote, Donna Haraway briefly asked "What if Western evolutionary and ecological sciences had been developed from the start within Buddhist instead of Protestant ways of worlding?" (2016, 176). Or thinking with the script, how might incorporating Theravada Buddhist understandings of "nature," "life," and "time" that are centred on impermanence, interconnectedness, and the non-self affect space exploration practices? Indeed, a small but growing number of theorists have asked how Buddhist ethics and care might be applied off-Earth (Capper 2023), as well as what Buddhist approaches to space exploration and the search for extraterrestrial intelligence might entail (Traphagan and Traphagan 2015).

I am not proposing the replacement of one ontology with another, nor suggesting that Theravada Buddhist worlds provide a singular solution to challenging normative logics. Rather, I advocate for thinking with a general metaphysics characterised by impermanence, decentredness, interdependence, and constant emergence. New worlds continually unfold, shaped by the conditions and components that are in motion, including our situated positions and common assumptions. An interdependent yet fragmentary metaphysics affords an ethnographic practice that engages with multiplicity by perpetually reconfiguring various elements.

While it is not possible to simply step outside our situated positions and accompanying assumptions, generating worlds through reconfiguring components is one way of destabilising seemingly set logics and revealing new potentials, which is a necessity for both anthropologists and astronomers who seek to make sense of unfamiliar realities (Houdart and Jungen 2015; Savransky 2021). This metaphysics can be applied as a tool for ethnographic methods, exemplified in this chapter by the co-creative workshop that brought together participants who typically would not collaborate to envision an imagined future. The same principles can also be applied theoretically, as evident in the reading of concepts like existence, nature, and selfhood through Theravada Buddhist conventions. Moreover, this metaphysics extends to the practical aspects of writing. By bundling various field sites (such as NARIT and UFO Kaokala), and methods (including observations, filmmaking, and co-creation), as well as recounting and interpreting field experiences back home in Berlin (at various desks, during different seasons overlaid onto my memories, photographs, and notes from Thailand), a new composition emerges—one among many possible compositions. After all, descriptions by researchers contribute to shaping worlds too (Jensen 2021). Through this open-ended practice, elemental parts can be tinkered with and reconfigured into new forms; forms that may be contested, or incomplete and inevitably, in the words of UFO Kaokala's Ann, "disperse again as particles in the cosmos."

Acknowledgements

Research for this chapter received funding from the Emslie Horniman Anthropological Scholarship Fund.

References

Aulino, Felicity. 2020. "From Karma to Sin: A Kaleidoscopic Theory of Mind and Christian Experience in Northern Thailand." *Journal of the Royal Anthropological Institute* 26: 28–44. https://doi.org/10.1111/1467-9655.13239

Aulino, Felicity. 2022. "Believe It and/or Not: Opening Up to Ontological Pluralities in Northern Thailand." *Anthropological Theory* 22 (2): 222–246. https://doi.org/10.1177/1463496221081321

Barker, Joshua. 2005. "Engineers and Political Dreams: Indonesia in the Satellite Age." *Current Anthropology* 46 (5). https://doi.org/10.1086/432652
Battaglia, Debborah. 2012. "Coming in at an Unusual Angle: Exo-Surprise and the Fieldworking Cosmonaut." *Anthropological Quarterly* 85 (4): 1089–1106. www.jstor.org/stable/41857291
Bauman, Whitney. 2015. "Religion, Science, and Globalization: Beyond Comparative Approaches." *Zygon* 50 (2): 389–402. https://doi.org/10.1111/zygo.12170
Blaser, Mario. 2016. "Is Another Cosmopolitics Possible?" *Cultural Anthropology* 31 (4): 545–570. https://doi.org/10.14506/ca31.4.05
Buchli, Victor. 2021. "Extraterrestrial Methods Towards an Ethnography of the ISS." In *Lineages and Advancements in Material Culture Studies: Perspectives from UCL Anthropology*, edited by Timothy Carroll, Antonia Walford, and Shireen Walton, 17–32. London and New York: Routledge.
Capper, Daniel. 2023. *Buddhist Ecological Protection of Space: A Guide for Sustainable Off-Earth Travel.* Lanham, MD, Boulder, CO, New York and London: Lexington Books.
Cassaniti, Julia. 2015. *Living Buddhism: Mind, Self, and Emotion in a Thai Community*. Ithaca, NY: Cornell University Press.
Casumbal-Salazar, Iokepa. 2017. "A Fictive Kinship: Making 'Modernity,' 'Ancient Hawaiians,' and the Telescopes on Mauna Kea." *Native American and Indigenous Studies* 4 (2): 1–30. https://doi.org/10.5749/natiindistudj.4.2.0001
Chakrabarty, Dipesh. 2000. *Provincializing Europe: Postcolonial Thought and Historical Difference*. Princeton, NJ: Princeton University Press.
Chuensamnaun, Wassana. 2014. *Extraterrestrial Talks About the "Golden Age"*. Published by the author.
Clifford, James and George E. Marcus. 1986. *Writing Culture: The Poetics and Politics of Ethnography*. Berkeley and Los Angeles: University of California Press.
de la Cadena, Marisol. 2015. *Earth Beings: Ecologies of Practice across Andean Worlds*. Durham, NC: Duke University Press.
Descola, Philippe. 2013. *Beyond Nature and Culture*. Translated by Janet Lloyd. Chicago, IL: University of Chicago Press. First published 2005.
Escobar, Arturo. 2020. *Pluriversal Politics: The Real and the Possible*. Translated by David Frye. Durham, NC: Duke University Press.
Gärdebo, John, Agata Marzecova and Scott Gabrial. 2017. "The Orbital Technosphere: The Provision of Meaning and Matter by Satellites." *The Anthropocene Review* 4 (1): 44–52. https://doi.org/10.1177/2053019617696106
Geertz, Armin W. 1984. "A Reed Pierced the Sky: Hopi Indian Cosmography on Third Mesa, Arizona." *Numen* 31 (2): 216–241. https://doi.org/10.2307/3269955.
Gubrium Aline and Krista Harper (eds). 2013. *Participatory Visual and Digital Methods*. Walnut Creek, CA: Left Coast Press.
Haraway, Donna. 1988. "Situated Knowledges. The Science Question in Feminism and the Privilege of Partial Perspective." *Feminist Studies* 14 (3): 575–599. https://doi.org/10.2307/3178066

Haraway, Donna. 1997. *Modest_Witness@Second_Millennium. FemaleMan©_Meets_OncoMouse™: Feminism and Technoscience*. New York: Routledge.

Haraway, Donna. 2016. *Staying with the Trouble: Making Kin in the Chthulucene*. Durham, NC: Duke University Press.

Harding, Sandra. 2008. *Sciences from Below: Feminisms. Postcolonialities, and Modernities*. Durham, NC: Duke University Press.

Holbrook, Jarita C., Rodney Medupe, and Johnson O. Urama (eds). 2008. *African Cultural Astronomy: Current Archaeoastronomy and Ethnoastronomy Research in Africa. Astrophysics and Space Science Proceedings*. Berlin: Springer.

Houdart, Sophie and Christine Jungen. 2015. "Cosmos Connections." *Gradhiva. Revue D'anthropologie et D'histoire des Arts* 22: 4–23. https://doi.org/10.4000/gradhiva.3016

Jackson, Mark. 2018. *Coloniality, Ontology, and the Question of the Posthuman*. London: Routledge.

Jasanoff, Sheila and Sang-Hyun Kim (eds). 2015. *Dreamscapes of Modernity: Sociotechnical Imaginaries and the Fabrication of Power*. Chicago, IL and London: University of Chicago Press.

Jensen, Casper Bruun. 2021. "Practical Ontologies Redux." *Berliner Blätter* 84: 93–104.

Kirsch, Thomas. 1977. "Complexity in the Thai Religious System: An Interpretation." *Journal of Asian Studies* 36 (2): 241–266. https://doi.org/10.2307/2053722

Kitiarsa, Pattana. 2005. "Beyond Syncretism: Hybridization of Popular Religion in Contemporary Thailand." *Journal of Southeast Asian Studies* 36 (3): 461–487. https://doi.org/10.1017/S0022463405000251

Knorr-Cetina, Karin D. 1981. *The Manufacture of Knowledge: An Essay on the Constructivist and Contextual Nature of Science*. New York: Pergamon Press https://doi.org/10.1016/C2009-0-09537-3h

Kohn, Eduardo. 2013. *How Forests Think: Toward an Anthropology Beyond the Human*. Berkeley: University of California Press.

Kuhn, Thomas S. 1962. *The Structure of Scientific Revolutions*. Chicago, IL: University of Chicago Press.

Lambek, Michael. 2013. "The Continuous and Discontinuous Person: Two Dimensions of Ethical Life." *Journal of the Royal Anthropological Institute* 19 (4): 837–858. https://doi.org/10.1111/1467-9655.12073

Larkin, Brian. 2013. "The Politics and Poetics of Infrastructure." *Annual Review of Anthropology* 42: 327–343. https://doi.org/10.1146/annurev-anthro-092412-155522

Latour, Bruno. 1993 [1991]. *We Have Never Been Modern*. Translated by Catherine Porter. Cambridge, MA: Harvard University Press.

Latour, Bruno. 2005. "'Thou Shall Not Freeze-Frame', or, How Not to Misunderstand the Science and Religion Debate." In *Science, Religion, and the Human Experience*, edited by James D. Proctor, 27–48. Oxford Scholarship Online. https://doi.org/10.1093/0195175328.003.0003

Latour, Bruno, and Steve Woolgar. 1986. *Laboratory Life: The Construction of Scientific Facts*. Princeton, NJ: Princeton University Press. First published 1979.

Law, John. 2015. "What's Wrong with a One-World World?" *Distinktion: Scandinavian Journal of Social Theory* 16 (1): 126–139. http://dx.doi.org/10.1080/1600910X.2015.1020066

Lempert, William. 2014. "Decolonizing Encounters of the Third Kind: Alternative Futuring in Native Science Fiction Film." *Visual Anthropology Review* 30 (2): 164–176.

Mair, Jonathan. 2012. "Cultures of Belief". *Anthropological Theory* 12 (4): 448–466. https://doi.org/10.1177/1463499612469588

McMahan, David, and Erik Braun (eds). 2017. *Meditation, Buddhism, and Science*. New York: Oxford University Press.

Mirmalek, Zara. 2020. *Making Time on Mars*. Cambridge, MA: MIT Press.

Omura, Keiichi, Grant Jun Otsuki, Shiho Satsuka, and Atsuro Morita. 2019. *The World Multiple: The Quotidian Politics of Knowing and Generating Entangled Worlds*. London and New York: Routledge.

Pickering, Andrew. 2017. "The Ontological Turn: Taking Different Worlds Seriously." *Social Analysis* 61 (2): 134–50. https://doi.org/10.3167/sa.2017.610209

Pink, Sarah. 2015. *Doing Sensory Ethnography*. London: Sage Publishing.

Raepeth, Somjit. 2014. *How to Use Extraterrestrial Technologies*. Published by the author.

Ratanakul, Pinit. 2007. "The Dynamics of Tradition and Change in Theravada Buddhism." *Journal of Religion and Culture* 1 (1): 233–257.

Rees, Tobias. 2018. *After Ethnos*. Durham, NC: Duke University Press.

Reid, Mark J. and Xing-Wu Zheng. 2020. "A New Map of the Milky Way." *Scientific American*. Online. Accessed 28 July 2023. www.scientificamerican.com/article/gaias-multi-billion-star-map-of-the-milky-way-keeps-getting-better/

Rubenstein, Mary Jane. 2014. *Worlds without End: The Many Lives of the Multiverse*. New York and Chichester, England: Columbia University Press.

Ruggles, Clive L. N., and Nicholas J. Saunders (eds). 1993. *Astronomies and Cultures*. Niwot: University Press of Colorado.

Russell, Catherine. 1999. *Experimental Ethnography: The Work of Film in the Age of Video*. Durham, NC: Duke University Press. https://doi.org/10.2307/j.ctv11cw7bq

Savransky, Martin. 2021. *Around the Day in Eighty Worlds: Politics of the Pluriverse*. Durham, NC and London: Duke University Press.

Sjöberg, Johannes and Alexandra D'Onofrio. 2020. "Moving Global Horizons: Imagining Selfhood, Mobility and Futurities Through Creative Practice in Ethnographic Research." *Culture & Psychology* 26 (4): 732–748. https://doi.org/10.1177/1354067X20922141

Smith, Rachel Charlotte, Kasper Tang Vangkilde, Mette Gislev Kjærsgaard, Ton Otto, Joachim Halse, and Thomas Binder (eds). 2016. *Design Anthropological Futures*. London and New York: Bloomsbury Academic.

Spivak, Gayatri Chakravorty. 1988. *Can the Subaltern Speak?* Basingstoke: Macmillan.

Ssorin-Chaikov, Nikolai. 2013. "Ethnographic Conceptualism: An Introduction." *Laboratorium* 5 (2): 5–18.

Stengers, Isabelle. 2010. *Cosmopolitics I*. Translated by Robert Bononno. Minneapolis: University of Minnesota Press.

Stodulka, Thomas. 2023. "Reflexivity, Engagement, Decoloniality: Shifting Emergences of Ethnography and Collaboration." In *In Tandem – Pathways towards a Postcolonial Anthropology | Im Tandem – Wege zu einer postkolonialen Ethnologie*, edited by Mirjam Lücking, Anna Meiser, Ingo Rohrer, 103–125. Springer Link.

Stonington, Scott D. 2020. "Karma Masters: The Ethical Wound, Hauntological Choreography, and Complex Personhood in Thailand." *American Anthropologist* 122 (4): 759–770. https://doi.org/10.1111/aman.13464

Strathern, Marilyn. 1988. *The Gender of the Gift: Problems with Women and Problems with Society in Melanesia*. Berkeley: University of California Press.

Suhr, Christian and Rane Willerslev. 2012. "Can Film Show the Invisible? The Work of Montage in Ethnographic Filmmaking." *Current Anthropology* 53: 282–301.

Swanner, Leandra. 2017. "Instruments of Science or Conquest? Neocolonialism and Modern American Astronomy." *Historical Studies in the Natural Sciences* 47 (3): 293–319. https://doi.org/10.1525/hsns.2017.47.3.293

Terwiel, Barend J. 1976. "A Model for the Study of Thai Buddhism." *The Journal of Asian Studies*, 35 (3): 391–403. https://doi.org/10.2307/2053271

Thongcharoen, Ann, Ploy Buranasi, and Lauren Reid. 2019. *The Key*. Unpublished script.

Traphagan, John. 2020. "Religion, Science, and Space Exploration from a Non-Western Perspective." *Religions* 11 (8): 397. https://doi.org/10.3390/rel11080397

Traphagan, John and Julian Traphagan. 2015. "SETI in Non-Western Perspective." In *The Impact of Discovering Life Beyond Earth*, edited by Steven J. Dick, 299–307. Cambridge: Cambridge University Press. https://doi.org/10.1017/CBO9781316272480.024

Valentine, David. 2017. "Gravity Fixes Habituating to the Human on Mars and Island Three." *HAU: Journal of Ethnographic Theory* 7 (3): 185–209 https://doi.org/10.14318/hau7.3.012

Verran, Helen. 2014. "Working with Those Who Think Otherwise." *Common Knowledge* 20 (3): 527–539. https://doi.org/10.1215/0961754X-2733075

Viveiros de Castro, Eduardo. 2004. "Perspectival Anthropology and the Method of Controlled Equivocation." Tipití: Journal of the Society for the Anthropology of Lowland South America 2 (1): 3–22. Available at: http://digitalcommons.trinity.edu/tipiti/vol2/iss1/1

Viveiros de Castro, Eduardo. 2014 [2009]. *Cannibal Metaphysics: For a Post-Structural Anthropology*. Translated by Peter Skafish. Minneapolis, MN: Univocal Publishing.

Wagner, Roy. 2016 (1975). *The Invention of Culture*. Chicago, IL and London: University of Chicago Press.

11 Divining South Korea's space age through Korean shamanism and astrology (*Myŏngni*)

Hae-Seo Kim

On 5 August 2022, South Korea launched a moon satellite "Danuri"[1] through a SpaceX Falcon 9 rocket from Cape Canaveral in Florida, in the US. After the launch of satellite Danuri, South Korea became the seventh country in the world, after the USSR, the US, Japan, the European Union, China, and India, to successfully launch a moon exploration satellite. As the nation celebrated the successful launch of the satellite and the beginning of what news media called the "Korean Space Age," there ensued a controversy about whether South Korea's President Yoon was relying on a shaman advisor to make important executive decisions through shamanistic divination. This chapter examines the South Korean "space age" as one that is simultaneously shaped by scientific and shamanistic/astrological relations, both at the national and local scales. This chapter is based on ethnographies and interviews with shamans and astrology readers in Seoul, Goheung, and Daejeon – three cities that have been shaped by the presence of the South Korean space station[2] and space research agencies.

The first part of this chapter provides a brief overview of Korean astrology and shamanistic divination, where they overlap, and where they are distinct. The second part of the chapter follows President Yoon's ambitions in outer space as well as his public engagement with Korean shamanism. Through interviews with astrology readers and shamans who engage with President Yoon's chart, this section analyses the co-constitution of scientific and shamanistic cosmologies in a public figure that shapes discourse and policy at the national scale. Shamanism and astrology have a complicated relationship to South Korean modernity, in which they have been stigmatised as "anti-science" or "backwards." This chapter historically contextualises the meaning of the South Korean president's public display of shamanism alongside his endorsement of the space program. The last part of this chapter is based on ethnographies with shamans and astrology readers in Goheung and Daejeon, where the nation's first space station and the nation's major scientific research

DOI: 10.4324/9781003437956-11

institutes, such as the Korean Aerospace Research Institute (KARI), are located. It examines how shamans and astrology readers help their clients navigate economic and social changes in their hometowns brought on by the space program and policies and help them face uncertainties in the future through divination.

Building on studies of outer space that centre multiple cosmological approaches and heterogeneous ways people relate to outer space (Valentine 2012; Lempert 2014; Casumbal-Salazar 2019; Hobart 2019; Maile 2021; Determann 2021), this chapter ethnographically traces how shamanistic and space scientific cosmologies are shown to co-constitute the material and social relations that make up South Korea's "Space Age." Methodologically, this chapter illustrates that an ethnography of Korean shamanistic and astrological divination is an ethnography of the socio-political environment in which outer space is explored in South Korea.

Brief overview of Korean astrology (*Myŏngni*)

In Korean astrology, *Myŏngni* or *Saju*, as it is more commonly referred to, spatial bodies such as the sun, moon, and the seven stars of the Big Dipper constellation, have held significant symbolic place in devotional and divination practices. In Korean astrology, the placement of Mars (*Hwasŏng*), Mercury (*Susŏng*), Venus (*Moksŏng*), Jupiter (*Kŭmsŏng*), and Saturn (*T'osŏng*) and their respective relationship to and combination with the five elements of nature (fire, water, wood, metal, and earth) shape the temperament, aptitude, interests, and ultimately the fortune of an individual (see Figure 11.1). A Korean cultural studies scholar, Yong-Hun Cho, writes that in *Myŏngni*, "the moment the umbilical cord is cut, the energy of the cosmos enters your body [...] *Myŏngni* narrates the influence of the seven planets on a person's life" (Cho 2011: 363).

Originating from China around 800 CE., different traditions of astrology are popular in China, Korea, and Japan. The Korean tradition of *Myŏngni* is practised throughout South Korea, and its attendant business of fortune telling was reportedly a $3.7 billion industry in 2018 (*The Economist* 2018). Korean astrology is most often performed by shamans. However, Korean astrology is also studied and practised widely by people who are not shamans, such as academics from various disciplines, bankers, and even Buddhist monks. While Korean astrology had historically been associated with shamans and was judged as sorcery and superstition, recently many Korean scholars in the humanities and social sciences have been reclaiming Korean astrology as an academic discipline (Ko 2012; Kang 2015; Yang 2020). These scholars distinguish *Myŏngni* from shamanism to show that it is a scientific, historically proven method of understanding events and

時	日	月	年
정재	主	편재	정관
辛	丙	庚	癸
卯	午	申	卯
정인	겁재	편재	정인
목욕	제왕	병	목욕

Figure 11.1 Astrology chart of a person born on 16 August 2023.

the psychology of human beings. In this chapter, I focus on the reading of Korean astrology, which is sometimes done by shamans and sometimes by other astrology readers. Therefore, it is important to keep in mind that while Korean astrology and shamanism are closely connected, they are not necessarily synonymous with each other.

Whether as an academic discipline or as a spiritual practice, for centuries Korean astrology has been an important way for people to relate to nature, the cosmos, and outer space, as well as a source of guidance for interpreting events and making decisions. Whereas Korean shamanism has been widely researched in relation to the country's political and socioeconomic changes, including with regards to the introduction of the Internet and cyber culture (Kim 2003; Kendall 2009; Yun 2019; Safarti 2021; Lee 2022), the role of Korean astrology is a relatively less-studied area, including within Korean studies and the anthropology of Korean shamanism (Kim 2019). Through ethnographic fieldwork with Korean astrology readers and shamans, this research aims to foreground the underexplored role of *Myŏngni* and illuminate how South Koreans understand their lives and futures through relating with outer space and planetary bodies in motion.

President Yoon's space age

In March 2022, Yoon Suk Yeol was elected as South Korea's twentieth president. Since then, Yoon has been at the forefront of pushing for South Korea's investment in outer space. At the same time, he has also had a widely publicised relationship with shamanism. Yoon faced much public criticism in the fall of 2021 when he showed up to a presidential candidates' broadcast debate with a vivid black tattoo on his palm that read "king." It was rumoured that a shaman had advised him to wear this temporary tattoo, which signified the powers needed to become the supreme ruler of the country (YTN 2021). Many shamans have since attested to the fact that such a symbol gives the wearer luck when making public speeches or public appearances.

Several opposition party political leaders have commented on Yoon's public display of a shamanist symbol. The opposition Liberal Party spokesperson has said "There is suspicion that the sign 'king' carries a shamanist meaning. This reminds us of the former President Park and Choi Soon-Sil. The public remembers what happens when an incompetent leader relies on superstition and magic to make a political decision" (Paek 2021). This is a reference to former President Park, who was impeached because of incompetence and her reliance on shamans for making political decisions. Another opposition party leader, Song, has said that "the Korean people have already made a judgement upon the leaders that have used magic to manipulate the people. The president is a person who should volunteer for the people and work for the people – it is ridiculous for Yoon to think of the presidential position as that of a 'king'" (Kwak 2021).

The public was offended that Yoon could compare the presidential position to that of a "king," or something given to him by cosmic powers, when the nation prides itself on being a liberal democracy. They were also offended at the very public display of his shamanist practices. While shamanism is widely practised throughout South Korea, it is often done in private and kept as an almost shameful secret (Kim 2003). This "cultural paradox" (Kim 2003) seems to also be gradually changing, with many celebrities and politicians openly seeking shamans and discussing it on YouTube and other social media. There are hundreds of online blog posts and YouTube videos by astrologers and shamans who discuss the charts of President Yoon and other public figures and their actions and decisions in the light of their astrological chart. In a recent talk show on YouTube hosted by former Secretary of State Park Ji-Won, a Korean astrologer says of Yoon:

> He has a lot of *sanggwan* in his chart. *Sanggwan* means to fight against established order. *Sanggwan* is not afraid of previous order. It fights against it [...] But a national leader needs *insŏng*, which is about learning and the ability to tolerate. Yoon doesn't have much of that. He doesn't listen to others much. This is intensified by his birth date, which symbolises the first star of the Big Dipper. It's very strong and it means people with this sign like to be the boss. It is very stubborn and proud. Do we need a president like that? We will see in time.[3]

I asked another Korean astrology reader, my interlocutor, if it is because of his chart (Figure 11.2) that Yoon is so invested in both South Korea's outer space program and shamanism. She says that Yoon's *Sanggwan* might indeed shape his interest in and fascination with the otherworldly, whether that's with astronomy, or with shamanism and astrology:

時	日	月	年
상관	主	편인	비견
癸	庚	戊	庚
未	午	子	子
정인	정관	상관	상관
관대	목욕	사	사

Figure 11.2 President Yoon Suk-Yeol's Korean astrology chart. Time of birth not exact.

> *Sanggwan* has been hated on historically in Korean astrology because it fights against established power and order. This was a complete taboo in hierarchical Confucian Korean society. But nowadays, *Sanggwan* is welcomed as it can stand up against power and injustice, and express itself artistically or through speaking such as through social media. *Sanggwan* can also signify that someone has spiritual and philosophical inclinations and is not afraid to express them.

She laughs and adds, "It also seems like because of his *Sanggwan*, he doesn't care what other people think. Everyone lives according to their astrology chart."

In a recent speech after touring the Goddard Space Center in Maryland together with US Vice President Kamala Harris, President Yoon told a nostalgic story about how he became enchanted with outer space:

> It still resonates deeply when on July 27th Neil Armstrong became the first human to set foot on the moon. I still vividly remember watching that historic moment on TV. It was during summer break when I was in third grade, and ever since for me personally, space has been the embodiment of dreams and new challenges and my visit today to NASA where all these dreams and challenges begin. This reminds me of that moment and has made me renew my resolve to never stop pursuing my dreams over the past six decades … Space has inspired and instilled the can-do spirit and countless people around the world by championing mankind's endeavours to explore the unknown universe … I've always believed that mankind's future lies in space.[4]

More than any previous South Korean president, Yoon has been a major advocate for more investment in South Korea's outer space research and development. During President Yoon's tour, NASA Deputy Administrator Pam Melroy and South Korea's Ministry of Science and ICT Minister Lee Jong-ho signed a joint statement of intent between the US and South Korea that promises further cooperation in science and space exploration.

One of Yoon's election promises was to "plant the South Korean flag on Mars in 2045." To achieve this, he proposed to make a new Korean bureau of outer space affairs, or what he proposed to call KASA (Korean Aeronautics and Space Administration). However, Yoon's unilateral decision to build KASA in the city of Sacheon in Gyeongnam province, in the south-east of the Korean peninsula, provoked a lot of resistance in other cities. He justified his decision through the rhetoric of "equal development across the nation (*Kyunhyŏnggaebal*)," but this decision was not welcomed by politicians and some residents of the regions such as Goheung and Daejeon, where there is an existing presence of the space program. They became defensive of their newly established positions as centres of South Korean space development and research.

Historically, Gyeongnam region, where the new KASA is to be located, has received much more governmental support for economic and industrial development than Goheung, which to this day remains an underdeveloped fishing and farming area and which welcomed the economic stimulus initiated by the arrival of the space industry. Gyeongnam's development is understood to have received preferential treatment by successive governments because the authoritarian dictators of the 1960s–80s originated from Gyeongnam region. As a result, Gyeongnam tends to have a more conservative voter base than Goheung region, which many people think is the real reason behind Yoon's decision to promote Gyeongnam as another centre of outer space in South Korea.

Others speculate that there may have been other reasons for Yoon's plans for Sacheon to host KASA. One of Yoon's close shaman counsellors, Chun Gong, has recently visited Sacheon, which raised questions among some of my shaman interlocutors about whether he was involved in some shamanic divination of where the nation's centre of the space program should be based. One of my astrology reader interlocutors said that she would not be surprised if "Chun Gong had told the President that Sacheon has good Feng Sui or is the right place to build South Korea's space program for it to thrive, which may in fact be correct, but may be ridiculous for most people to accept." She continued, "Who knows, maybe it is Chun Gong who is pushing for Yoon to develop South Korea's space program – maybe he sees a bright future in it." Chun Gong has previously caused controversy by being allegedly involved in moving the presidential office to a different part of the capital. It was alleged that Chun Gong had provided counsel to President Yoon about Feng Sui and astrological luck with regards to making the move, which caused much anger and resentment among the public.

These controversies raise interesting questions about the place of astrological readings and shamanism in "divining" the nation's future on Earth and in outer space. The public's negative reaction to the president's close relationship to shamans and astrology readers, while at the same time relying on astrology readers and shamans themselves, reveals the nation's "paradoxical" (Kim 2003) relationship to different cosmologies – scientific and astrological. The nation's relationship to Korean astrology, shamanism, and science goes back to its first encounters with colonial modernity. The Japanese colonial government (1910–45) viewed Korea's Indigenous shamanist and divinatory practices as superstitious, backwards, and antithetical to modernity. The Japanese colonial government used Korea's shamanism and fortunetelling practices to show the superiority of Japanese culture and to justify Japanese colonial rule of Korea. Since then, Korea's leaders' vision of postcolonial nation-building has centred around science and technology (Kim 2015), and shamanism, Korean astrology, and other related practices have continued to be attacked as backwards and even criminalised (Yi and Won 2005; Kim 2009). Although the persecution of shamans abated in the 1990s as they became accepted as emblems of "Indigenous Korean culture," shamanism and Korean astrology are still widely stigmatised, even among those who practise them.

Banu Subramaniam has shown how the allegedly separate zones of the secular and the religious are in fact co-constituted in Indian modernity (Subramaniam 2019, 27). Similarly, shamanistic, astrological, and space scientific cosmologies not only coexist but are co-constituted in South Korean political modernity, as shown in the speculations over President Yoon's use of astrological advice to make policy decisions about where to move the presidential residence, and even where to place South Korea's space bureau. The co-constitution of these cosmologies can also be seen in the proliferation of astrologers' and shamans' reading of the president's chart to interpret his policy decisions through the lens of shamanism and astrology.

Divination in space towns

Park, a woman in her early sixties, is a shaman and Korean astrologer in Goheung, an area composed of 400 small islands in the southernmost part of the Korean peninsula. Over the years, she has seen a lot of changes in her hometown, especially since 2009, with the construction of the nation's first space station in Naro Island, off the coast of Goheung. While Goheung has historically been a fishing town, its politicians and businesspeople have been repositioning it as a new centre of outer space development in South Korea. Numerous space museums and space-themed hotels have been built, and construction of a space theme park is underway. Some of the roads have been re-named as "*Wuju-ro*," which means "Space Road," or "Let's go to Space" depending on how it is translated. By branding itself as the nation's first "Space Town," Goheung is trying to attract space tourists who come to see the launch of satellites and space shuttles, as well as attracting educational tours from schools around South Korea. Shaman Park says, "I think Goheung became a much better place to live in after the space station, because the roads got better, and more people started paying attention to us."

While Shaman Park supports the space station, she has expressed scepticism about the efficacy of the scientific method in approaching outer space:

> I think the space program is contributing to Goheung, but I don't think that the scientists understand the cosmos. We shamans connect with the whole of the cosmos, the stars, and nature. Shamans bring the gods into our bodies and worship them. We are connected. But I don't think the scientists are connected to the cosmos.

"The government should not shift space industries to Sacheon," Park says when I ask her about President Yoon's plans to move the centre of the space program to Sacheon. While she has scepticism about scientists' ability to decipher the meaning of outer space, she still supports the presence of the space program in Goheung. She hopes that the space program will "attract young men and women to move out here and start families." With the migration of young people to bigger cities since the nation started rapidly industrialising in the 1970s and '80s, Goheung has been facing an ageing and declining population, like many other rural parts of South Korea. Park says it is bad for shamanic and divination business as well:

> In the past, Goheung was a thriving fishing town where young people stayed because they could make good money. So, there were a lot of young people here, making families and such. People used to seek shamans to hold *kut* [an elaborate shamanist ritual] for fish catch. Shamans used to have more young clients who wanted to seek advice for marriage, love, children, and so on. Now, all the young people want to move to bigger cities like Seoul and Pusan. All that remains in Goheung are old people (*Nŭlgŭni*) like me. They don't have that many questions about life, and so they don't seek shamans out much. It's bad for the shamans.

Shaman Park sees the space station and related industries as a renewed opportunity for Goheung to attract young professionals to move to the area and make it their home, which would in turn be good for the divination business.

A few days later, I visit her again while she is reading one of her clients' son's fortune. The female client, like many eager Korean parents, wants to know if her son will get into a university in Seoul, which is where most of the nation's top universities are located. The son is in high school and wants to become a scientist, and she says that he wants to become an aerospace engineer, as he grew up inspired by the space station being built in his hometown. She emphasises that he is a good student, but he is nervous about getting accepted into one of the top universities in Seoul. He had to enrol at a high school with specialisms in maths and science in a different part of the country, because the educational system in Goheung is lacking.

I sit down while Shaman Park writes down the boy's chart: his birth year, month, date, and time. She shakes her ceremonial bell to consult the spirits, and contemplates for a second before she answers, "Yes, spirits say he can go to a university in Seoul." The client, Mrs. Lee, is much relieved to hear this. Park continues talking while looking at the child's chart: "He has much luck with his *Kwan*, which is his work sign or institutional sign. He has a lot of *In*, which is a learning sign, coming into his chart, so he will be able to study all he wants and get the positions he desires." While Park does not necessarily believe in what scientists do with what she calls their "satellite business," she wishes for successful young people to continue to come back to live in Goheung, and she sees the growing space-related industries as a chance to attract them. Her endorsement of the child's bright future in the sciences is an endorsement of her vision of rebuilding Goheung as an attractive place to live in.

Mrs Lee says "Thank you" many times to the shaman, gives her some cash, and turns to me and invites me out for coffee. As we head to a café in downtown Goheung, Mrs Lee tells me that she wants her son to become a successful scientist and go abroad to work for "international" (*kukchejŏk*) agencies like NASA. She thinks that there are limited opportunities for her son in South Korea and wants her son to keep his options open in other countries. "The [South Korean] government doesn't pay its scientists enough, and we all know it," she exclaims. She emphasises that she does not want her son to stay in Goheung in the future: "We spent a lot of money on our son, so that he could attend the specialised high school in Kwangju. We want to support him as much as we can, so that he is not restricted to live in this rural town."

While Mrs. Lee has certain ambitions and expectations for her son, she is worried that President Yoon will divert the space industry to a different region. She is hoping that the space program in Goheung will continue to grow, attracting tourists and educated personnel, such as scientists and engineers, to the region. President Yoon's plans to shift the centre of the space industry to a different region in South Korea was making her and other supporters of the space program anxious about Goheung's future as the hub of space exploration in South Korea.

Since the inauguration of President Yoon, stock market investment for private space industries in South Korea have increased. To take advantage of this momentum, around seventy South Korean private space companies released a statement to support the government's policies for space development and the "realisation of a space economy." While start-ups and companies have been taking advantage of the government's support for outer space development, residents of areas that used to be at the centre of the "space economy" in South Korea have been expressing increasing uncertainty about the future of their hometowns. An interlocutor from Daejeon, a city two hours south of Seoul, said that the residents of Daejeon worry that if the centre of science research and development moves away from Daejeon, real estate prices will fall.

Since the 1960s, Daejeon has benefited from its proximity to Seoul, and has hosted South Korea's premier scientific agencies and educational institutions, such as the Korean Advanced Institute of Science and Technology (KAIST) and the Korean Aerospace Research Institute (KARI). In a country where science, technology, and education are very much at the centre of the economy, shifting some of the city's infrastructure to another part of the country is adamantly opposed by its residents, politicians, businesspeople, and scientists (Jin, Jung, and Kim, 2022). Many scientists and politicians in Daejeon have been upset by the recent presidential decision to move the space research facility to a different part of the country, and some of them seek shamans to ask for guidance.

"Many clients come to ask questions about money, health, real estate, and education. And people these days are very anxious, so they come see me, or other shamans and astrologers," a shaman and astrology reader based in Daejeon tells me. "People want to know if it's a good time to invest, if real estate prices will fall, or if their child will get employed." Since she operates very close to the KAIST university, I asked her if professors or students at the university come to see her. She says mostly students come to ask about their grades, which major to pick, and when to graduate, but some scientists have also come to see her. When I ask her why she thinks scientists are opposed to moving the space industry to Sacheon, she says, "Well, it's because scientists are people too. They own homes and they care about the value of their apartments. Scientists also have children and families, and they care about their children's education."

What my shaman interlocutor is telling me is that scientists do not wish to move to a more rural part of Korea because many of them have children and most of the good schools, including those that specialise in maths and science, are in the larger cities like Daejeon, not in more rural cities like Sacheon or Goheung. Scientists also have apartments and families and care about the value of their homes. These "quotidian" material reasons, such as where to live and where to send their children to school, interact with national-level policy decisions, including where to place the national space bureau. In both Goheung and Daejeon, residents are uncertain about their futures. They consult astrology readers and shamans to gain certainty about the future and to make sound decisions about education, investment, and moving. Sang Yeong Kim, a retired financial planner turned Korean astrology reader, says in an interview with a Korean newspaper that

> People worry about the success and failure of what they do. The advantage of Saju [Korean astrology] [...] is that success and failure can be measured through the axis of time. That is, if someone wants to row a boat, we can predict through someone's astrological chart when the water is going to come and tell them when they should do what they want to do.
>
> (Park 2022)

According to my astrology readers and shaman interlocutors, reading the Korean astrological chart is like astronomy in that it is a way of understanding the components of the cosmos and the placement of the Earth and other planets in the cosmos. The ability of Korean astrologers and shamans to provide counsel through understanding the placement of celestial bodies in the cosmos and to guide people through one's life path and temporal decision making is what draws people to Korean astrology and divination. At the same time, this ethnography demonstrates how shamanistic and astrological divinations in Goheung and Daejeon are situated materially and socially in relations created by the space program and related infrastructures. Space scientific infrastructures and policies impact the reading and interpretation of Korean astrology and shamanism. Shamans and astrology readers, in turn, guide their clients – eager parents, scientists, and anxious investors – through changes brought on by the space program and its infrastructures. Following the consultations of shamans in space towns and examining divination based on individual charts illustrates the co-constitution by the scientific and shamanistic cosmologies, by which Korean "Space Age" is being made, imagined, and lived.

Conclusion

Korean astrologer and writer Kang Hun, in *Myŏngni: Reading Fate* (Kang 2015), writes that the study of Korean astrology "transform[s] the study of the skies, or astronomy into the study of humans, or the humanities." This research brings together the social study of outer space and the study of human society, or anthropology, through an ethnographic focus on Korean astrology and shamanism. At the national level, President Yoon of South Korea has been very publicly rumoured to make important state decisions relying on the advice of a shaman. At the local level, South Korean people similarly navigate economic and social changes brought on by the state policies on outer space by seeing their shamans and asking them for advice. In this way, shamanistic and scientific cosmologies co-constitute the material and social relations that make up South Korea's "Space Age."

Methodologically, this research places outer space as simultaneously acting upon and embedded in people's lives, whether that's through the nation's science and technology policies, through the president's relationship to both shamanism and outer space, or through people relying on shamans and astrology readers to manage the economic and social changes in their lives. This research therefore moves away from the study of outer space as a place "out there" that is to be explored, developed, and even colonised. Rather, it contextualises outer space as it is lived in South Korean people's everyday lives, embedded in South Korea's economy, politics, and history (Messeri 2016; Olson 2018). Furthermore, the historicising of the relationship between science and astrology in South Korea illustrates how science is also co-produced by infrastructures, environments, and different cosmologies (Jasanoff 2004; Subramaniam 2019). Understanding how multiplicity of cultures and peoples engage space "can provide another blueprint for how we might engage with space beyond Earth" (Milligan 2023; Smiles 2020; Praet 2023). A focus on how outer space is understood and lived in multiple ways in different cultural contexts will help de-centre Euro-American histories of outer space exploration and diversify our understandings of space exploration and human futures in space.

Note on Romanisation and Translation

I use the McCune–Reischauer romanisation system to transliterate Korean words into Latin script. Personal names in Korean follow the cultural convention of family name followed by given name, except for individuals who publish under or who prefer Western conventions. Place names in South Korea follow the South Korean revised romanisation system, which was adopted by the South Korean government in 2000. I forgo romanisation for Korean-language bibliographic sources but include the English-language translation. Unless otherwise indicated, all translations from Korean interviews and bibliographic sources are my own.

Acknowledgements

I would like to thank the Social Science Research Council's International Dissertation Research Fellowship (SSRC-IDRF) and Wenner-Gren Foundation's Dissertation Fieldwork Grant for supporting my fieldwork research. I would also like to thank the editors of this volume – Jenia

Gorbanenko, Adryon Kozel, and David Jeevendrampillai – for their tremendous help and insight throughout the editing process.

Notes

1 The Korean name of the satellite "Danuri" comes from *Dal*, which means the moon, and *Nuri*, which means the cosmos. "Danuri" therefore means moon's cosmos, or the cosmos through the moon. The name signals South Korea's ambition to get closer to the cosmos through the moon.
2 In this chapter, "space station" refers to South Korea's satellite launch pad and space research centre in Goheung, South Korea, not the International Space Station.
3 Medici Media. "Will I become King? Four Commonalities in the Astrology of Presidents." YouTube, uploaded by Medici Media, 12 January 2023, www.youtube.com/watch?v=K-_huFZ6rcs.
4 Vice President Harris and South Korean President Remarks at Goddard Space Center, 25 April 2023.

References

Casumbal-Salazar, Iokepa. 2019. "'Where Are Your Sacred Temples?' Notes on the Struggle for Mauna a Wākea." In *Detours: A Decolonial Guide to Hawai'i*, edited by Hōkūlani K. Aikau and Vernadette Vicuña Gonzalez. Durham, NC: Duke University Press.

Cho, Yong-Hun. 2011. "Korea's Astrology." In idem, *The Joy of Korean Studies*. Seoul: Humanist. 조용헌. 2011. "한국의 역학." *한국학의 즐거움*. 서울: 휴머니스트.

Determann, Jörg Matthias. 2021. *Islam, Science Fiction, and Extraterrestrial Life: The Culture of Astrobiology in the Muslim World*. New York: I.B. Tauris.

Hobart, Hi'ilei Julia. 2019. "At Home on the Mauna: Ecological Violence and the Fantasies of Terra Nullius on Maunakea's Summit." *Native American and Indigenous Studies* 6(2): 30–50.

The Economist. 2018. "In South Korea fortune-telling will soon be a $3.7 bn business." 24 February. www.economist.com/asia/2018/02/24/in-south-korea-fortune-telling-will-soon-be-a-37bn-business

Jasanoff, Sheila. 2004. *States of Knowledge: The Co-Production of Science and Social Order*. London: Routledge.

Jin, Nayeon, In-Seon Jung, and Dong-Hee Kim. 2022. "Daejeon resists the decision to place the Korea's Space Agency in Sacheon, Gyeongnam." 28 April. *Daejeon Ilbo*.진나연, 정인선, 김동희 기자. "항공우주청 경남 사천 설치 결정에 대전 지역 반발 확산". www.daejonilbo.com/news/articleView.html?idxno=1516800

Kang, Hun. 2015. *Myŏngni: Reading Fate*. Seoul: DolBaeGae. 강헌. *명리: 운명을 읽다*. 돌베개.

Kendall, Laurel. 2009. *Shamans, Nostalgias, and the IMF: South Korean Popular Religion in Motion*. Honolulu: University of Hawai'i Press.

Kim, Chongho. 2003. *Korean Shamanism: The Cultural Paradox*. Burlington, VT: Ashgate.

Kim, David. 2019. "Divination and its Potential Futures: Sensation, Scripts, and the Virtual in South Korean Eight-Character Astrology." *Material Religion: The Journal of Objects*, Art, and Belief 15: 599–618.
Kim, Sang-Hyun. 2015. "Social Movements and Contested Sociotechnical Imaginaries in South Korea." In *Dreamscapes of Modernity: Sociotechnical Imaginaries and the Fabrication of Power*, edited by Sheila Jasanoff and Sang-Hyun Kim. Chicago, IL: University of Chicago Press.
Kim, Seong Nae. 2003. "Korean Shamanic Heritage in Cyber Culture." *Bibliotheca Shamanistica* 11: 279–295.
Kim, Young Mi. 2009. *Their New Village Movement*. Seoul: Green History Publishers. 김영미 그들의 새마을 운동. 푸른역사.
Ko, Mi-Sook. 2012. *How to Use My Fate: Saju Myŏngni and Anti-Oedipus*. Seoul: BookDramang. 고미숙. 2012. *나의 운명* 사용설명서*: 사주명리학과 안티 오이디푸스*. 북드라망.
Kwak, Eun-San. 2021. "Yoon Suk-Yeol's Controversy with the 'King' on his Palms…Speculation of Shaman's Involvement." *Syegye Ilbo*. 3 October. 곽은산. "윤석열 '손바닥 王자 논란…무속인개입설." 2021.10.03. https://n.news.naver.com/mnews/article/022/0003624824?sid=100
Lee, Joonseong. 2022. *The Assemblage of Korean Shamanism: Mediatization and Territorialization*. Switzerland: Springer Nature.
Lempert, William. 2014. "Decolonizing Encounters of the Third Kind: Alternative Futuring in Native Science Fiction Film." *Visual Anthropology Review* 30(2): 164–176.
Maile, David Uahikeaikalei'ohu. 2021. "On Being Late: Cruising Mauna Kea and Unsettling Technoscientific Conquest in Hawai'i." *American Indian Culture and Research Journal* 45(1): 95–121.
Medici Media. 2023. "Will I become King? Four Commonalities in the Astrology of Presidents." YouTube, uploaded by Medici Media, 12 January 12· www.youtube.com/watch?v=K-_huFZ6rcs.
Messeri, Lisa. 2016. *Placing Outer Space: An Earthly Ethnography of Other Worlds*. Durham, NC: Duke University Press.
Milligan, Tony. 2023. "From the Sky to the Ground: Indigenous Peoples in an Age of Space Expansion." Space Policy 63. https://doi.org/10.1016/j.spacepol.2022.101520
Olson, Valerie. 2018. *Into the Extreme: U.S. Environmental Systems and Politics Beyond Earth*. Minneapolis: University of Minnesota Press.
Paek, Sang-Jin. 2021. "Yoon Suk-Yeol's 'King' Makes Him a Superstitious Magician President…Yoon Claims it was from his Supporter." *Kookmin Ilbo*. 4 October. 백상진. "윤석열 '王' 자에 "미신, 주술 대통령"… 尹 '지지자 응원 메시지'" 2021. 10.04. https://n.news.naver.com/mnews/article/005/0001474390?sid=100
Park, Soo Ho. 2022. "Financial Planner Turned Astrology Reader: Asked About South Korea's National Future." *Maeil KyungJae*. 박수호. 2022. "증권사 임원에서 변신한 역술인…추석맞아대한민국국운물어봤더니." www.mk.co.kr/economy/view.php?sc=50000001&year=2022&no=804384

Praet, Istvan. 2023. "Anthropology and Contemporary Space Exploration, with a Note on Hopi Ladders." In *The Routledge Handbook of Social Studies of Outer Space*, edited by Juan Francisco Salazar and Alice Gorman, 57–70. London: Routledge.

Safarti, Liora. 2021. *Contemporary Korean Shamanism: From Ritual to Digital.* Bloomington: Indiana University Press.

Smiles, Deondre. 2020. "The Settler Logics of (Outer) Space." Essay in Society and Space, 26 October. www.societyandspace.org/articles/the-settler-logics-of-outer-space

Subramaniam, Banu. 2019. *Holy Science: The Biopolitics of Hindu Nationalism.* Seattle: University of Washington Press.

Valentine, David. 2012. "Exit Strategy: Profit, Cosmology, and the Future of Humans in Space." *Anthropological Quarterly* 85(4): 1045–1067.

Yang, Chang-Soon. 2020. Myŏngni Psychology: When Life doesn't go as Wanted. Seoul: Dasan Books. 양창순. 2020.명리심리학: *사는게 내 마음 같지 않을 때*. 다산북스.

Yi, Yong-Bhum, and Kang Jeong Won. 2005. "Korean Folk Beliefs and Science." *Asian Comparative Folklore* 28(2): 83–104. 이용범, 강정원. "민속신앙과 과학." 비교민속학 28(2): 83–104.

YTN. 2021. "Letter 'King' in the palm of Yoon Suk Yeol ... 'Back to the Time of Choi Sun Sil?' Yoon faces criticism." YouTube, uploaded by YTN on 2 October. www.youtube.com/watch?v=UyYRrgBwNas

Yun, Kyoim. 2019. *The Shaman's Wages: Trading in Ritual on Cheju Island.* Seattle: University of Washington Press.

12 Space intentionally left blank

Peter Timko and Karlijn Korpershoek

According to the Fédération Aéronautique Internationale, space is a place. It begins at the Kármán line, an invisible, largely arbitrary border 100 km above the Earth's mean sea level (FAI 2018). This conception of space—as a location, out there, beyond a specific altitude—is only one definition. Space also exists down on Earth as a social sphere or social category. Different individuals and groups will have different conceptions of what space is, what it is for, and what it means to be in it. In this way, space in practice is a multiple, a shifting, unsettled concept that can be contorted and warped at the behest of context and convenience. In the past decade, as various projects of space expansion have proliferated and intensified, anthropologists and fellow researchers have grappled with "the radically different and multiple natures of space" (Battaglia, Valentine, and Olson 2015, 247). Some have explored the way the strange affordances and novel dynamics of extraterrestrial life can upend and reconfigure established conceptual frames and material relationships. Others still have identified the way that the pursuit of off-Earth engagement has precipitated new social relations on the ground, both within the tight-knit groups of professionals (Vertesi 2012), and at the larger scales of international astropolitics (Wang 2009).

The two authors of this chapter are part of the ARIES Project (Anthropological Research into the Imaginaries and Exploration of Space), a research group based at Jagiellonian University that contributes to and expands on this growing body of literature by tasking a team of anthropologists to approach space through an ethnographic lens (see also Ojani, Chapter 1 of this volume). Our respective projects investigate the ways relations between space and communities are changing in different contexts around the world. Karlijn explores the entanglement of state power, colonial histories, and local identity at launch sites in French Guiana, while Peter investigates the social dynamics emerging as space operations increasingly move beyond direct

DOI: 10.4324/9781003437956-12

state management. The indeterminacy of space emerges as a common thread running through our otherwise dissimilar field sites. It invites an extended examination of what space actually means across different contexts and communities.

Grappling with the shifting multiplicities of space calls for making nimble methodological changes to our research activities, particularly in how we construct space as an object of study in our fields. As scholars who have come before us have often pointed out, a certain elusiveness is inherent to ethnographic research: "No matter how much one tries, the field always seems to slip away, seep in, expand, and transgress the spatial and temporal bounds in which we attempt to contain it" (Hussain 2021, 145). In the following, we reflect on how our own methodological approaches and conceptual frames have adapted to better serve the protean qualities of our research foci and how this process contributes to more detailed, grounded research outcomes.

The chapter proceeds as follows: each of us in turn presents a short narrative detailing how our respective approach to space evolved through our fieldwork. First, Peter discusses broadening his inquiry into the new space economy in Los Angeles; then, Karlijn unpacks the vagaries of space as it emerges in Kourou, French Guiana. We conclude with a discussion of the parallels that emerge between our work and reflect on the ways our projects complicate prevailing understandings of space even where it may seem most settled. By laying out our own stories, we hope to show how flexibility becomes an asset when looking into the slippery terrain of space and how sometimes the best approach to understanding a mutable topic is to eschew rigidity and embrace multiplicity.

Peter: The expanding universe of new space

In my original research proposal, submitted along with my application to Jagiellonian University, I wrote that I wanted to uncover the various socio-technical imaginaries animating "new space," the then-common term for the commercial space industry. At the time, the agenda seemed straightforward enough: contact a few headline-grabbing new space companies, hang around as they build giant machines bound for orbit, and figure out what forms of imaginative framework scaffold their work (Jasanoff and Kim 2015). In my head, the project would be analogous to other in-depth examinations of technology-intensive workplaces, such as Gusterson's *Nuclear Rites* (1996), which carefully probed the internal lives of the people engaged in building nuclear weapons at Lawrence Livermore National Laboratory in California. Unfortunately, this initial plan floundered almost immediately upon launch.

I put "new space" in quotations here deliberately, for, as I found, this term points to an ideal more than a reality. In colloquial use, the phrase remains focused on large, private companies like SpaceX, though in actuality, the industry is much larger. The early months of my PhD were spent poring over articles and industry reports in an attempt to zero in on the ideal "new space" firms to approach. In this survey, I looked into the origins of the first space race and its fallout, explored the histories of individual space companies, and read dozens of taxonomies slicing up space operations into various categories—upstream, downstream, launch, earth observation, ground segment. The possible groupings were endless. On top of this, I conducted informal interviews with numerous people whose work fed the project of space expansion. I spoke with engineers, planetary scientists, public relations specialists, and welders, among others. We spoke about their career paths, their daily lives, and their opinions on which companies were going to make it and which were bound for dissolution. The goal was to cobble together a consensus picture of "new space," a rough description of which companies and activities are included and deserve attention. In other words, I wanted to "make the cut," to carve out a silhouette of new space from the "seamless reality" of the larger world (Candea 2009).

In the end, all this probing was enlightening, but not in the way it was originally intended to be. Where I sought clarity, I found complexity. "New space" proved to be a shifty concept with no clear borders. Every "new" operation was deeply entangled with "old space," through similar flows of technologies, personnel, and funding—and, as NASA chief economist Alexander MacDonald points out in *The Long Space Age*, these connections have deep roots (2017). Today, it is difficult to parse the business of space as a discrete economic sphere. Instead, activity in space cuts across economic sectors ranging from agriculture and insurance to mining and pharmaceuticals—"it is horizontal," as one early interlocutor put it. It became increasingly clear that focusing on "new space" as a circumscribed field or singular, coherent movement made little sense. Pinning down a definition to ground future fieldwork was not just unnecessary, it was likely obscuring what was actually going on. Looking back, I am reminded of Durkheim's concept of "prenotions," which I encountered through Desmond's writing on relational ethnography (2014). These ready-made ideas hang "as a veil interposed between the things and ourselves," and "resembling ghost-like creatures, distort the true appearance of things, but which we nevertheless mistake for the things themselves" (2014, 31). The idealised form of "new space," which could somehow be held up as a foil, separate from other space capacities or terrestrial activities, was such prenotion. I needed to look at the messy tangle outside the quotes.

Following this, the project shifted from a tight focus on specific businesses to the wider social and economic dynamics animating contemporary space expansion, an ongoing process I began seeing as the new space economy. This new framing is more expansive, drawing on both ecosystems' thinking, which looks at the "indirect, complex, and non-linear relationships among actors" in economic fields (Orlova, Nogueira, and Chimenti 2020, 2), as well Gibson-Graham's notion of diverse economies, which calls for an "economical approach" that includes a "wider range of social relations [that] bear on economic practices, including [...] trust, care, sharing, reciprocity, cooperation, divestiture, future orientation, collective agreement, coercion, bondage, thrift, guilt, love, community pressure" (2014, 151). This looser frame opened the aperture of my research to include a more panoramic view.

Social scientists have often advanced wariness of "methodological containerism" in various forms, questioning the primacy of preordained units as a basis of analysis—a critique that has been applied at scales ranging from nation states to museums (Macdonald, Gerbich, and Von Oswald 2018). Along these lines, anthropologists Urban and Koh (2013) have argued the merits of research that finds connections and blurs distinctions between the internal life of corporations and the ways they may impinge on the world around them. Picking up these themes means looking beyond the cloistered dynamics of a specific space company to follow the ways the space economy met with and related to various people, communities, and spaces. As shown below, the merits of this less rigid approach were clear: it allowed room to see the broader context—concepts and connections that would otherwise be cropped out the frame. Of course, this still involves making "a cut," a curatorial practice that can yield vastly different findings depending on what is brought into focus (Schacter 2020). In the fall of 2022, I set off to conduct fieldwork on site in Los Angeles, California with this in mind, ready to let the frame dilate and distort, accommodating additional facets of the space economy. In the following example, I show how this more exploratory, curatorial approach allowed for some of the most important findings in my research.

Tracing connections in the new space economy

Southern California is home to one of the most productive aerospace industries in the world. In recent years, numerous privately owned space companies have set up operations here, especially in the South Bay region of Los Angeles. I began my fieldwork speaking with workers at one of these companies, a small launch services provider. My contacts there were young—in their 30s or below. One, an avionics engineer, pointed out that he was recruited directly from his university. This was part of the company's early growth strategy; the founders made outreach visits to university engineering departments and rocket clubs. As the difficulty of sourcing labour had been a recurring theme in my preliminary interviews, it seemed that this relationship between launch companies and student rocket clubs was an under-examined facet of the space economy's mechanics.

There are multiple student rocket clubs throughout Los Angeles; the highest profile may be the University of Southern California Rocket Propulsion Lab (USC RPL). Founded in 2005, the club is staffed entirely by students. Each year, members design, build, and launch a rocket—notably, in 2019, the RPL's Traveler IV was the first fully student-designed and -built vehicle to cross the Kármán line. This club, with its volunteer membership and largely educational mission, falls outside most common visions of "new space" as seen in the popular press. Industry surveys written up and shared by consulting firms and space business boosters are just as likely to leave it out. Yet, the RPL and similar organisations are integral to the space economy's development. Spending time with the club at their workshop south of downtown and speaking with current members gave insight into how this relationship works.

There is a clear and ongoing reciprocity between the club and the growing private launch sector: many members join RPL to develop the abilities necessary to get a foothold in the industry; for instance, when asked about future plans, several students told me they were aiming for gigs at SpaceX. Students use their club experience, particularly leadership roles on various projects, as a means to land interviews, internships, and eventually jobs. The club's reputation carries weight, and the existing alumni network can provide useful connections as well. Meanwhile, students interning at companies bring back technical skills and insider industry know-how. These insights, such as how to organise a project or approach a problem, are incorporated into the Lab's operations and culture, making the RPL an even better training ground for future launch company workers. The result of this dynamic is a solid pipeline connecting graduates to jobs and companies with a tailor-made labour pool. It is no surprise then that some of the RPL's backers are companies which may benefit from the arrangement, such as SpaceX, Blue Origin, and Relativity Space (a manufacturer of 3D printed rockets, founded by RPL alum Tim Ellis).

It is well established that universities are crucial in the production of skilled workers for technical fields. However, investigating how this relationship works with student rocket clubs fleshes out a more nuanced picture of how the space economy is developing. One subtlety that was revealed talking with RPL is a very self-conscious awareness that they are part of the space industry, particularly as a training ground for its future workers. One way this manifests is that the members are overtly concerned with workplace composition and culture. While aerospace engineering has traditionally been dominated by certain demographics, members of the RPL pointed out that they desire to build a club inclusive in terms of race, gender, and sexuality by, for example, making a conscious effort to mentor marginalised students or actively calling out hostile behaviours. By shaping their own lab environment, they hope to encourage a broader range of people to participate and instil a certain set of values in future members. This functions as a form of "anticipation work," meant to shape the course of the future space economy to better fit their desires (Steinhardt and Jackson 2015). RPL members know that their future colleagues will come from clubs like theirs; thus by proactively fostering a diverse membership and inclusive norms, they are shaping a future industry more in line with their values.

This is just one thread that binds university rocket clubs to the growing private launch sector, and just one interrelationship among many throughout the wider space economy (Johnson, Chapter 7 of this volume). How this process will influence the future remains to be seen. It seems likely that as these fresh relations coalesce, they will shape who goes to space, what happens there, and how this process changes life on Earth. In any case, a different methodological cut or curation may not have brought it into view. As mentioned, Los Angeles and its environs already has a long history with the aerospace industry. Our understanding of the sector and its impacts has benefited from scholars exploring the aerospace sector's own vagaries, contingent entanglements, and blurry edges (Westwick 2012). As more activity turns toward the stars, it is necessary to bring the same broad, exploratory disposition to examining the space economy as it develops and stakes out its own footprint in southern California and around the globe. Keeping my fieldwork tightly bound within a static idea of "new space" may have entirely missed the emerging dynamics discussed above—for this, I am glad to have left the space open.

Karlijn: The state of "outer space"

While Peter's work focuses on the realm of "new space," my research led me to a location that has been closely associated with the space industry for a significant period: French Guiana. Its history has been extensively documented in one of the earliest ethnographers of outer space, Peter Redfield, who in *Space in the Tropics* (2000) and elsewhere describes "a setting where a routine form of rocketry directly crosses the remains of less final frontiers" (2002, 813). I am particularly attracted to the tension between the framing of French Guiana as the "ultra-periphery" and "outermost" region of the European Union in political discourse (Boatcă 2018; Hoefte, Bishop, and Clegg 2015) on one hand, and as central to the European space industry (ESA 2022a, 2022b) on the other hand as its sole operational heavyweight launch site. That is why in late 2023, I took off to French Guiana to explore what "outer space" means in Kourou, and specifically, what meanings it may hold beyond those sustained within the formal industry.

It is tempting to assume that finding "outer space" is straightforward in a self-proclaimed *ville spatiale* (space town) that welcomes you with a mock-up rocket and a clearly time-worn billboard declaring you have entered *la ville d'avenir et reussite* (the town of the future and success). However, soon into the research, it became clear that "outer space" in French Guiana is complex, multifaceted, and often elusive. The methodological challenge became how to capture both the omnipresence and ambivalence of the space industry to the people who live here. For many of my interlocutors, "outer space" feels alien, distant, or simply, uninteresting, while for others it appears in configurations that align with, are born out of, or completely contradict those offered by the spaceport.

Leaving the edges of "outer space" undefined allows for the research to open in unexpected directions and to acknowledge the complicated relationship between communities in the area and the space industry, where what is supposedly "out there" is seeping through to the local (Beery 2016). In this section, I highlight three significant ways in which it does: firstly, through the enmeshment of land rights and space exploration; secondly, through the infrastructure of Kourou; and lastly, through different initiatives that provide alternatives to the dominant space narrative.

Expropriation in French Guiana

In December 2023, I spoke to a teacher at the local high school, who explained: "Kourou is not about space, it is about the grounds/Earth (my translation—*terre* in French) used to get there." Looking at how terrestrial borders collide with atmospheric ones (Silver 2023) provides insights into the ways that the European space industry is entangled with lingering colonial remains. In 1946, French Guiana was politically integrated into France as an overseas department. At the time, the region was still a penal colony, created almost a century earlier and closed only in the 1950s (Renneville 2007; Sanchez 2018). After the closure, the French government found an alternative use for the territory, albeit one that drew on similar notions of "remoteness" and "emptiness" (Gorman 2007). With Algerian independence in 1962 spelling the end of their rocket testing site in Hammaguir, the French began looking for a new location. Two years later, French Guiana was chosen from a list of several possibilities as the most optimal replacement.

In addition to the launch site, the French Space Agency (CNES) also built the infrastructure in the surrounding region of Kourou (Chambaz et al. 2021). European space agency administrators arrived with their own expectations for the kinds of social services required to support their workforce who also came from mainland Europe. To build the infrastructure that they deemed necessary, the existing networks and communities were completely disregarded. Hundreds of people living in the region were displaced, rehoused, and forced to change their way of life, as their nomadic agricultural practices were made impossible.

The history of expropriation attached to the launch site is exemplary of the way in which "outer space" is not out there, but deeply entangled with land rights issues on Earth (Gorman 2007). The decision to force the Creole families living in the region to move out of their large, wooden houses into small concrete apartments is a difficult and often neglected topic in Kourou. There are efforts to bring more attention to this history. In 2023, I spoke to the author of *Mémoires des expropriés de Kourou et Sinnamary* (Memories of the Expropriated of Kourou and Sinnamary, Chocho-Dufail 2023). Through her writing and speaking at schools across French Guiana, she hopes to better inform the younger generations about the past. In its own way, the CNES is also working towards shedding more light on this history: the space museum, currently closed for renovation, will have a section dedicated to the expropriated families in the revamped exhibition hall. What this revival of attention to these colonial practices will provoke remains to be seen in the coming years. However, what is clear is that "outer space" does not exist as separate from Earthly power practices in French Guiana, so leaving the research open to talk about land rights was essential to grasp the different directions in which the industry's power is exerted.

Infrastructure of Kourou

The archives of the space centre hold historical documents that detail the processes that led to the creation of the launch site and Kourou. Access to these records during my fieldwork served as a pivotal method to comprehend the discrepancies between outward presentation and internal decision-making (Barry 2015) by stakeholders in the launch site. They document how the infrastructure of Kourou not only resulted in the expropriation, but also in an urban layout that mimics the hierarchical power dynamics of the base. For the most senior administrators, large villas were built along the beach. Building outwards, the rank of employees decreased, and the size of the houses shrank.

This opened up the infrastructure of Kourou as a method in itself. Not only does a focus on the urban infrastructure reveal the lingering "half-life" of empire as posited by Redfield (2002), but the infrastructure itself is a way of resistance (Larkin 2013). In 2017, the launch site was occupied by a group of protesters demanding more investment for French Guiana. The rocket base became a point of leverage and caused the suspension of an Ariane 5 launch until an agreement was reached for more money for the overseas territory, ending the demonstrations (Clark 2017; Breeden 2017). Close to twenty years earlier, the closure of the road running alongside the launch site, now known as "*route de l'espace*" (space road) caused uproar in the local community. The rerouting of the main road to make literal space for the launch site significantly increased the commute time for residents. This infrastructural change came to symbolise the continuing force exerted by the French government and space agency, and local elects argued that "France only develops in French Guiana that which interests it, notably the base of Kourou" (Redfield 2002, 806; see also Osbourne, Chapter 5, and Chinigò and Nieber, Chapter 4 of this volume, for further discussion of colonialism and space infrastructure).

The more I engaged with this topic, the more I became aware of how Kourou's infrastructure catered specifically for other people like me, and interrogating this became part of my approach. I grew conscious that coming from mainland Europe for a temporary stay and doing (at least some) work at the CSG, I am part of the key target group that the CNES had in mind while putting together the infrastructure of the city. How I use the infrastructures, social services, and entertainment, not just as part of "active" fieldwork, but specifically as part of navigating daily life, became a methodological asset. "Outer space" became personally very close; not as a location "out there" but as the way of living here on Earth as imagined by the space industry.

Myriad of other meanings of space

There are several efforts, affiliated to a varying extent with the space industry, which are trying to widen what "outer space" means in the French Guianese context. For instance, *Station K* is a Fablab in the centre of Kourou. This educational facility, modelled after a concept created at the media lab of MIT, provides a collective infrastructure to work with a variety of devices, such as 3D printers and VR goggles. The facility is run by a former ESA engineer and created at the initiative of, among others, the CNES. *Guyane Astronomie* is another recent attempt at making "outer space" less hegemonically European in French Guiana. Set up after the launch of the James Webb Space Telescope, it promotes astronomy practices throughout the overseas territory. Their current focus is on developing ethno-astronomy by creating a better and shared understanding of the relation with the stars of the several Amerindian communities throughout the Amazon. Finally, *Amazonie Spatiale* was a decentralised writing residency that brought together writers, scientists, and artists and delivered two projects: the collaborative publication of an imagined "lost book" and *The Amazonautes Journal*, a collection of residency participants' journals. These different projects show how "outer space" can exist outside of the confines of the rocket base, albeit important to note that often they share funders.

Leaving space blank allows for these different configurations to arise from how "outer space" is experienced, practised, and/or imposed on people. The methodological challenge of studying "outer space" is met by looking around, behind, and forward, as well as looking up.

Leaving space for multiplicities

As members of the ARIES team, space is the thread tying together both of our projects. At the same time, our actual research led us to vastly different locations, different people, and different questions. In this way, the multiplicity of space has always been apparent. Our encounters in the field have only made this clearer. Across both our field sites, the nature of space—what it is, means, does, could do, or could be—remained malleable and shifting. Embracing this indeterminacy proved to be a crucial step in deepening our inquiries. In French Guiana, Karlijn found that the conspicuous absence of outer space pushed her research away from an essentialising narrative where outer space overshadows other experiences in the territory. "Outer space" became the backdrop rather than the central aspect of much of the research. In Peter's work on the space economy, leaving the research design fluid allowed for different tendrils of the contemporary space industry to come to the fore. It opened a path to find the reciprocal relations between space companies and other, less visible groups—dynamics that may shape how the future might look.

The divergent paths our research took and the diverse notions of space we met with resonate with existing work oriented toward the cosmos. In the words of Fraser MacDonald, outer space exists as a "culturally configured site of knowledge and power where philosophical, scientific and aesthetic discourses intersect with socio-economic, technological and political forces" (2007, 594). The myriad spaces these arrangements produce do not always amiably align. As many have pointed out—and often critiqued—"outer space" can be constructed "as a frontier" (see Trevino 2020; Messeri 2017), or "yet another 'unknown' to be conquered and bent to America's will" (Smiles 2020, 2). It can be a site of liberatory potential, where Earthly injustice is discharged for new horizons (Triscott 2016), or a host of mirage utopias that will never come to pass (Tutton 2021). Space is also Sky Country, a place intimately connected to Aboriginal peoples and populated by their ancestors (Mitchell et al. 2020). Back on the terrestrial surface, disparate relations to space animate life paths, community action, and political controversy from the launch sites off Boca Chica (Szolucha 2023) and Spaceport America (Sammler and Lynch 2021) to the mountaintops in Thailand (Reid 2023, also Chapter 9 of this volume) and Mauna Kea, Hawaii (Maile 2021) and French Guiana.

On the surface, activity at our field sites would appear to be concomitant to firmly established understandings of space—broadly speaking, the launch facilities at the "*ville spatiale*" and the rocket factories of Los Angeles both emerge from a technoscientific tradition with antecedents stretching back to the original Space Race and earlier. Yet even within this paradigm, complexity can still be found, if one is open to finding it. By leaving space blank—conceptually, methodologically—our projects were more apt to meet nuance rather than polish it away or leave it to the side, unacknowledged and unexplored. With this approach at hand, we could better describe the social reality we encountered on-the-ground: that the multiplicity of space is the norm, not an aberrant phenomenon found at peripheries or at moments of exception. Assessed at a more intimate scale, even "conventional" space settings are made up of a complex mosaic of perspectives, processes, and, indeed, ways of seeing "outer space" itself (see Jeevendrampillai and Fortais, Chapter 7 of this volume)

Still, recognising this complexity and sitting with its tensions rather than seeking resolution is only one step in a longer enterprise. Our ethnographic projects must also be open to seeing these multiplicities on their own terms—to examine how they are made, where they meet, and the textures of these encounters. There is a temptation to place the presence of difference into pre-existing narratives, to make stories of conflict and competition. Observations from each of our field sites suggest, though, that other registers of relation are just as vital. The working conditions and immediate concerns at the RPL do not closely resonate with the discourses of "elite escape" identified in some private space projects. Still, the club and private industry actors maintain a close, reciprocal relationship, through which the student members hope to realise a future of their own making. Here, the undetermined contours of space are not an obstacle but an avenue for optimism. In French Guiana, there is far from a singular experience of space. For some, it represents the lingering coloniality of an imposed metropolitan project; for others, space is of no interest despite inhabiting a "*ville spatiale*." Others still see it as the cornerstone of Kourou's identity. By leaving space blank, the simultaneous distance and proximity to space, its ambiguity, and its ambivalence can all co-exist.

In some ways, space is more visible than ever before. Ambitious expansion projects regularly grab headlines, politicians posture about its geostrategic importance, and new technologies like the James Webb Space Telescope bring deep cosmic vistas down to Earth. Yet these same processes obscure other views of the cosmos—as dark skies groups warn, light pollution and LEO traffic threatens to disrupt and obscure the stars for millions around the globe. As social scientists, we know there is no single space story, and we should always attend to this complexity. As space activity continues to draw in and touch more and more people and communities, it is important these changes are put under critical scrutiny. This includes remaining open to the multiple, shifting perspectives which frame "outer space," ambitions towards it, and ideals projected through it, especially from people and groups not always included in the conversation. Space has never been empty, but that does not mean its future is already settled.

Acknowledgements

Research for this article received funding from the National Science Centre, Poland, project number 2020/38/E/HS3/00241.

References

Barry, Andrew. 2015. "The Oil Archives". In *Subterranean Estates*, edited by Hannah Appel, Arthur Mason, and Michael Watts. Ithaca, NY: Cornell University Press.

Battaglia, Debbora, David Valentine, and Valerie Olson. 2015. "Relational Space: An Earthly Installation". *Cultural Anthropology* 30 (2): 245–56.

Beery, Jason. 2016. "Unearthing Global Natures: Outer Space and Scalar Politics". *Political Geography* 55 (November): 92–101.

Boatcă, Manuela. 2018. "Caribbean Europe: Out of Sight, out of Mind?". In *Constructing the Pluriverse*, edited by Bernd Reiter, 197–219. Durham, NC: Duke University Press.

Breeden, Aurelie. 2017. "Strikes Shut Down French Guiana, With Effects Resonating in Paris". *The New York Times*. www.nytimes.com/2017/03/27/world/europe/french-guiana-general-strike-france.html.

Candea, Matei. 2009. "Arbitrary Locations: In Defence of the Bounded Field-Site". In *Multi-Sited Ethnography*, edited by Mark-Anthony Falzon, 24–47. Milton Park: Routledge.

Chambaz, Bernard, Bertrand Dezoteux, David Redon, and Gérard Azoulay. 2021. *Traversée Culturelle Dans Les Archives de l'espace. Vol.2: Le Centre Spatial Guyannais*. Paris: Observatoire de l'Espace/CNES.

Chocho-Dufail, Juliana. 2023. *Mémoires Des Expropriés de Kourou et Sinnamary*. Cayenne.

Clark, Stephen. 2017. "Operations Resume at European-Run Launch Base in French Guiana". *Spaceflight Now*. https://spaceflightnow.com/2017/04/25/operations-resume-at-european-run-launch-base-in-french-guiana/.

Desmond, Matthew. 2014. "Relational Ethnography". *Theory and Society* 43 (5): 547–79. https://doi.org/10.1007/s11186-014-9232-5.

ESA. 2022a. "ESA – ESA Council Meeting at Ministerial Level – Media Conference". www.esa.int/ESA_Multimedia/Videos/2022/11/ESA_Council_Meeting_at_Ministerial_Level_-_Media_Conference.

———. 2022b. *Space Transportation Development 'an Investment in Europe's Future'*. The European Space Agency.

FAI. 2018. "Statement about the Karman Line". www.fai.org/news/statement-about-karman-line.

Gibson-Graham, Julie Katherine. 2014. "Rethinking the Economy with Thick Description and Weak Theory". In *Current Anthropology*. Vol. 55. Chicago, IL: University of Chicago Press.

Gorman, Alice. 2007. "La Terre et l'Espace: Rockets, Prisons, Protests and Heritage in Australia and French Guiana". *Archaeologies* 3 (2): 153–68. https://doi.org/10.1007/S11759-007-9017-9/FIGURES/2.

Gusterson, Hugh. 1996. *Nuclear Rites: A Weapons Laboratory at the End of the Cold War*. Berkeley: University of California Press.

Hoefte, Rosemarijn, Matthew Bishop, and Peter Clegg. 2015. "Still Lonely After All These Years?: Contemporary Development in the 'Three Guianas' ". *Caribbean Studies* 43 (2): 83–113.

Hussain, Zahra. 2021. "Open Inquiry: Fielding the Field". In *Navigating the Field: Postgraduate Experiences in Social Research*, edited by Mildred Oiza Ajebon, Diego Astorga de Ita, and Yim Ming Connie Kwong, 145–57. New York: Springer.

Jasanoff, Sheila, and Sang-Hyun Kim. 2015. *Dreamscapes of Modernity: Sociotechnical Imaginaries and the Fabrication of Power*. Chicago,IL: University of Chicago Press.

Larkin, Brian. 2013. "The Politics and Poetics of Infrastructure". *Annual Review of Anthropology* 42: 327–43.

MacDonald, Alexander C. 2017. *The Long Space Age: The Economic Origins of Space Exploration from Colonial America to the Cold War*. New Haven, CT: Yale University Press.

MacDonald, Fraser. 2007. "Anti-Astropolitik — Outer Space and the Orbit of Geography". *Progress in Human Geography* 31 (5): 592–615.

MacDonald, Sharon, Christine Gerbich, and Margareta Von Oswald. 2018. "No Museum Is an Island: Ethnography beyond Methodological Containerism". *Museum and Society* 16 (2): 138–56.

Maile, David Uahikeaikalei'ohu. 2021. "On Being Late: Cruising Mauna Kea and Unsettling Technoscientific Conquest in Hawai'i". *American Indian Culture and Research Journal* 45 (1): 95–122.

Messeri, Lisa. 2017. "Why We Need to Stop Talking about Space as a Frontier." https://slate.com/technology/2017/03/why-we-need-to-stop-talking-about-space-as-a-frontier.html.

Mitchell, Audra, Sarah Wright, Sandie Suchet-Pearson, Kate Lloyd, Laklak Burarrwanga, Ritjilili Ganambarr, et al. 2020. "Dukarr Lakarama: Listening to Guwak, Talking Back to Space Colonization". *Political Geography* 81 (August): 102218.

Orlova, Alina, Roberto Nogueira, and Paula Chimenti. 2020. "The Present and Future of the Space Sector: A Business Ecosystem Approach". *Space Policy* 52 (May): 101374.

Redfield, Peter. 2000. *Space in the Tropics: From Convicts to Rockets in French Guiana*. 1st ed. Berkeley: University of California Press. www.ucpress.edu/book/9780520219854/space-in-the-tropics.

———. 2002. "The Half-Life of Empire in Outer Space". *Social Studies of Science* 32 (5–6): 791–825.

Reid, Lauren. 2023. "Divergent Extraterrestrial Worlds: Navigating Cosmo-Practices on Two Mountaintops in Thailand". In *The Routledge Handbook of Social Studies of Outer Space*, edited by Juan Fransisco Salazar and Alice Gorman. Milton Park: Routledge.

Renneville, Marc. 2007. "Les Bagnes Coloniaux: De l'utopie Au Risque Du Non-Lieu". Criminocorpus https://doi.org/10.4000/criminocorpus.173.

Sammler, Katherine G., and Casey R. Lynch. 2021. "Spaceport America: Contested Offworld Access and the Everyman Astronaut". *Geopolitics* 26 (3): 704–28.

Sanchez, Jean-Lucien. 2018. "Le Bagne Commence à La Libération". *HAL Open Science* 3: 58–63. https://shs.hal.science/halshs-01409137.

Schacter, R. 2020. "A Curatorial Methodology for Anthropology 1". In Lineages and Advancements in Material Culture Studies (pp. 190–204), edited by R. Schacter, T. Carroll, and A. Watford. Routledge.

Silver, Jake. 2023. "Tear Gas in Orbit: On the Olfactory and the Extraplanetary in Palestine". American Ethnologist, February.

Smiles, Deondre. 2020. "The Settler Logics of (Outer) Space". www.societyandspace.org/articles/the-settler-logics-of-outer-space.

Steinhardt, Stephanie B., and Steven J. Jackson. 2015. "Anticipation Work: Cultivating Vision in Collective Practice". In *CSCW 2015 – Proceedings of the 2015 ACM International Conference on Computer-Supported Cooperative Work and Social Computing*, 443–53. Association for Computing Machinery, Inc.

Szolucha, Anna. 2023. "SpaceX's Biggest Rocket Flies for the First Time: But Do We Understand What This Actually Means?" *The Geek Anthropologist*. https://thegeekanthropologist.com/2023/04/20/spacexs-biggest-rocket-flies-for-the-first-time-but-do-we-understand-what-this-actually-means/.

Trevino, Natalie B. 2020. "The Cosmos Is Not Finished". The University of Western Ontario. https://ir.lib.uwo.ca/etdhttps://ir.lib.uwo.ca/etd/7567.

Triscott, Nicole. 2016. "Transmissions from the Noosphere: Contemporary Art and Outer Space". In *The Palgrave Handbook of Society, Culture and Outer Space*, edited by James S. Ormrod and Peter Dickens, 414–44. London: Palgrave Macmillan.

Tutton, Richard. 2021. "Sociotechnical Imaginaries and Techno-Optimism: Examining Outer Space Utopias of Silicon Valley". *Science as Culture* 30 (3): 416–39.

Urban, Greg, and Kyung Nan Koh. 2013. "Ethnographic Research on Modern Business Corporations". *Annual Review of Anthropology* 42: 139–58.

Vertesi, Janet. 2012. "Seeing like a Rover: Visualization, Embodiment, and Interaction on the Mars Exploration Rover Mission". *Social Studies of Science* 42 (3): 393–414.

Wang, Sheng Chih. 2009. "The Making of New 'Space': Cases of Transatlantic Astropolitics". *Geopolitics* 14 (3): 433–61.

Westwick, Peter J. 2012. *Blue Sky Metropolis*. Berkeley: University of California Press.

13 Methodological sensibilities in outer spaces

Valerie Olson

In order to broaden the scope of their methodological approaches, ethnographers have been multiplying "multi"s for decades. These efforts recognize the inherent multiplicity of social space, time, and perception. Space is multi-sited; time is multi-temporal; relational perceptions are multi-scalar; the body is a body multiple; ethnographic analyses can be multimodal.

Underlying these multis, most of the time, lies an unquestioned singular. The taken-for-granted ground for Western social scientific sensibilities is the one-and-only Scientific Earth. The place where, scientists hold, human beings biologically live and evolve. So even multisitedness, multitemporality, and multiscalarity can be tacitly arrayed within a single Earthly dimension. By questioning that tacit unidimensional ground, the ethnographers in this volume convey their multidimensional sensibilities of space, time, and relations. I am struck by how those sensibilities—not the invention of new methods per se—connect these chapters across ethnographic topics and intentions.

When ethnographers include extraterrestrial spaces in fieldwork, they engage and participate in otherwise dimensions of lived experience. These are spaces that are deeply and broadly social, even if people do not have direct bodily experiences in them. The pieces in this volume illustrate how contemporary spaces of social belonging include other planetary futures and pasts, how everyday kinships between far-apart things like stars and wheat become naturalized, and how so-called traditional spiritual practices de-center the Scientific Earth in the service of human becoming. Going ethnographically multidimensional like this does not necessarily require new field methods in a formal empirical sense, but it invites us to *perceive differently*.

In what follows, I highlight how ethnographers in this volume cultivate *interpositional* and *contemplative* sensibilities. These multidimensional sensibilities override Aristotelian categories of noncontradiction. There is

DOI: 10.4324/9781003437956-13

no behind-the-scenes either/or or here/there imperative about being in one place or another. To be specific, ethnographers of outer spaces can't just act as if people are really only here on Earth and are not really in outer spaces as well. And they can't proceed from the assumption that their interlocutors are only imagining or speculating about their connections to outer spaces.

Instead, the authors here sense otherwise. They can adopt an interpositional sensibility, along with their interlocutors, that they *are simultaneously multiply positioned within* spatial relations that include but aren't limited to planet Earth. In addition, they can adapt a *contemplative* conjoined-body-mind sensibility that allows them *to actually perceive the simultaneously multiworldly experiences* of their interlocutors.

These multidimensional sensibilities are not ordered by scientific chains of being that start from the individual standing on the surface of Earth and go "up and down" mediated linear scales, from body down to cell to molecule to atom, from body to nation, from planet to universe. Adopting an interpositional or contemplative sensibility makes it possible to sense complex webs of earthly and outer spatial relations. It becomes possible to perceive how people feel the intimacies of stars and souls, of past Earth bodies and future Martian bodies, of earthly and galactic communities.

Such ways of perceiving human space, time, and relationality are not locked into the idea that there is a default "a real Earth baseline" of experience – either for interlocutors or for ethnographers. The very idea of a "base line" just ratifies linear empirical orderings of experience and, by extension, analysis. Multidimensional perceptual capacities are not bound to hierarchical Aristotelian model of non-contradictory and non-simultaneous relations of space, time, beings, and things.

It's not that ethnographers of outer spaces deploy such sensibilities and other ethnographers do not. But ethnographers of outer spaces can face criticism for following the connections that people make between earthly and nonearthly things. Colleagues can critique those projects as social scientifically irrelevant or inconsequential. Or worse, as unrealistic or fantastical. The question to ask is where such normative beliefs about an earthly vs. unearthly "reality" are coming from? The chapters in this volume address the experiences of social groups for whom such analytic criticism – and the single-point position it comes from – just isn't meaningful.

Going multidimensional

My interest in ethnographic multidimensionality is the result of doing outer space fieldwork and through years of assisting students and colleagues to dream up anthropological projects. I join other scholars of late who have begun to note the value of thinking multidimensionality about ethnographic work (see *inter alia*, Star 1999; Khurshid 2017; Thompson 2017; López Caballero 2021; Peterson and Olson 2024). In these instances, multidimensionality does not just *encompass* multisitedness, multitemporality, or multiscalarity. It is not bound to categories of kind or scale. Multidimensional project design and analysis encourages crisscrossing all kinds of meta-categorical boundaries.

When my colleague Kris Peterson and I help students design multidimensional projects, we encourage them to highlight conceptual connections that reflect social experiences rather than social scientific logics. They can then design projects that highlight how seemingly incommensurate social phenomena connect in the emergent dynamics of social life. Although many ethnographers only come to see such emergent connections after they end their projects, students can perceive and work with them from the get-go. They can explore links between the making of new Mormon souls in Peru and migration policy crises in the US state of Utah, between solar power startups and the rise of religious nationalism in India, between evangelical political activities and financial literacy movements in Appalachia, and between home security apparatuses and the ecological legacies of slavery in Jamaica (Peterson and Olson 2024). These projects blend topical dimensions in ways that encourage cross-topical conversations in fascinating ways.

In this mode of perception, ethnographers can follow actual earthly and outer spatial connections in their fieldwork. When, for example, they design projects that show how Mars is *actually ethnographically connected* to earthly political, social, and cultural worlds, they challenge the basic rules of ethnographic data collection and what counts as ethnographic area studies. They have to collect data about a spatial experience – such as in low earth orbit or a planet or interstellar space – without actually going there. They also have to proceed as if human extraterrestrial pasts and futures are actual and as if people are already relating to more-than-terran beings (Battaglia 2006; Lepselter 2016). Ethnographers of outer spaces end up refielding and reworking the basic definitions of being, knowing, materiality, life, kinship, and settler colonialism (Olson 2023; Messeri and Olson forthcoming). Doing such ethnographies *requires* unearthbound sensibilities.

Cultivating interpositional sensibilities

Positionality is a key meta-concept in ethnographic work. In the late twentieth century, ethnographers began to describe their interlocutors' social experiences as "positional" with respect to others in a broader social geography. Later, feminist ethnographers, ethnographers of color and queer ethnographers called for all ethnographers to identify their own positionality with respect to their interlocutors. This perspective holds that positional asymmetries in ethnographer-interlocutor interactions shape an ethnographer's analytic choices and perspectives.

In contemporary ethnographies, positionality is understood to be intersectional and almost always shifting as new forces transform social geographies. This alters how ethnographers relate to interlocutors in changing contexts, how they and interlocutors experience those changes differently, and how the positionality of persons and things shift as they move from place to place (see *inter alia*, Reynolds and Orellana 2009; Berry et al. 2017; Crossa 2015). To pay attention to positionality is to acknowledge that shifting contexts and processes are constantly impacting access to sovereignty, safety, resources, and mobility. Outer space studies have to deal with positionality – of others and of themselves – in unusually broadly scoped ways.

One of the defining dimensions of outer space studies is the way that space technologies, places, and resources re-shape Earthly social positionality structures. But for space ethnographers, these processes are often as much *in transition* as anchored in place. Ethnographers detailing work on other planets show how scientists' social positionalities on Earth extend to Mars but their experiences of embodiment – from sleep to their senses – are interactively "on" Earth and "on" Mars (Mirmalek 2020; Hoeppe 2012).

Other ethnographers have followed the actions of technoscientific colonial powers to build space science sites and create truth-claims about outer spaces. They show how outer spaces become intercolonial, linking them to colonized places on Earth and in ways that pre-territorialize them in the name of nations or companies (see *inter alia* Hobart 2019; Goodyear-Ka'ōpua 2017; Smiles 2020; Charbonneau 2021). In those sites, colonized peoples experience the cosmic interpositionality of colonialism. Scholars of outer spaces also point out that decolonization can be interpositional as well, indexing the possibilities of Black and Indigenous futures no longer constrained by the one and only Scientific Earth (Lempert 2014; Casumbal-Salazar 2017; Samatar 2017).

What makes such studies innovative is that ethnographers themselves are analyzing and writing from an interpositional stance, feeling themselves both here and there. This volume's authors looked for ways to project their ethnographic sensibilities into the extremely interstitial spaces of their interlocutors. In Chapter 2, Giles Bunch addresses positionality explicitly, showing how astronauts' extremely variable positionalities shed light on the complexity of technoscientific labor organization and embodied hierarchies *in extremis*. As laborers aboard the International Space Station, astronauts are extremely elite but also extremely sacrificial subjects. What these elite subjects allow to be done to them in space does not map onto class models of alienated labor. However, Bunch argues that this "special" case reflects new forms of hierarchy and power that result in the proliferation of new technoscientific contexts, such as, conceivably, the other spaces of virtual worlds and realities.

Makar Tereshin and Denis Sivkov in Chapter 6 take us into the 'Earth–Kosmos' museum's interspatially situated homeland. The team's methodology was interpositional along multiple dimensions. In order to attend to interlocutors' experiences of being intimately earthly and broadly cosmic, they "blurred" methodological and epistemological boundaries in order to "be in one place together or in several places at once, to experience the same event in different ways" (this volume, 97). This allows them to show that what it means to be "local" in these sites does not scale simply to the "global," but is relative to the temporally and spatially "universal."

Also based in a museum context, Alana Osbourne's chapter (Chapter 10) on the Greenwich Royal Observatory shows how museum goers are made to experience meteorites as "portals" for a colonial "matrix" without beginning or end (this volume, 85). This positions viewers in a universe made for extraction, making the prospect of geological resources *and* extraction infinite across all times and spaces. Osbourne puts herself into this interpositional space, showing how the displays create the overlapping "geo-logics" of science and imperialism (this volume, 88). Citing Yusoff et al. 2022, she shows how "imperial imaginings" can be multidimensional when they invoke interplanetary connections (this volume, 85).

In Chapter 7, as Anne W. Johnson connects people working in Mexican space science with artists working on Mars as a space intimately connected to the history of Mexico, she brings her readers into her interpositional experience of cultural spheres that interrelate Earth and Mars. As some of her interlocutors speculate about future Mexicans travelling to Mars, others position themselves as already there. She shows that Mars, even within one country, is a planet with multiple relational "milieux" that interconnect people, places, objects, times, and experiences. Johnson points to Kathleen Stewart's definition of "milieux" as "prismatic singularities," meaning that they are spatially and materially unique but also multiple (this volume, 121). Choosing to speak of Mars's milieux in the plural shows that Mars is, for many people, as socially multidimensional a world as Earth.

Other authors in the volume find new ways to describe their experience of social worlds that do not center the one and only Scientific Earth, using terms that invoke an interpositional felt sense. In Chapter 9, David (Jeeva) Jeevendrampillai and Sarah Fortais invite their readers to vacillate between a Martian and earthly reading of the ethnography of a space analogue mission. The term "vacillate" conveys both a sense of shifting between but also the hesitation of landing anywhere. Because, in this case, the "field" of a practice space mission is "landing" somewhere and nowhere at the same time, which is also true for the ethnographers. This means that the ethnographers have to "decouple" from notions of realness, boundedness, and linearity. The field is multidimensional because experience, for interlocutors and ethnographers, is multidimensional.

Peter Timko and Karlijn Korpershoek, in Chapter 12, intensify the need for an ethnographic interpositional sensibility in their work on the Earth/space divide. "Space" is multiple across all kinds of dimensions, even for people in government or industrial space work. For interlocutors in French Guiana and southern California, "outer space" is a shifty locale, conceptually, economically, and legally. People working in launch sites and rocket facilities deliberately cultivate that shiftiness, constantly keeping outer space a "blank space" in order to serve visions that serve economic speculation and national territory-making. This is not just another terra nullius because "blank space" could shift into something more defined at any time. Ethnographers of outer spaces must try to perceive interlocutors' interpositional powers, vulnerabilities, and aspirations, and somehow convey both the certainties and uncertainties that such positions generate.

As ethnographers cultivate interpositional sensibilities, they can be called into different kinds of internal *dis*positions. As they encounter interlocutors who are actively imagining and speculating about outer space ambitions or futures, ethnographers do not have to engage in those practices. However, as several of the authors in this volume show us, there is value in the practice of contemplating such possibilities as more than "just" possible. Cultivating a contemplative sensibility allows ethnographers to perceive outer space dreams and futures with their interlocutors.

Cultivating contemplative sensibilities

Ethnographers argue that social processes of imagination and speculation – by themselves and in their interlocutors – are not unreal. Instead, they are deeply and expansively transformative. Well-known cultural histories and ethnographies of imagination and speculation show, for example, how new nations come into being through textual imaginaries of community (Anderson 1983), how social categories are actualized as they are being imagined (Valentine 2007), how designers are actually actively bringing new forms to life when they imagine collaboratively (Murphy 2005), how people who run market stalls in Lagos are engaging and impacting acts of global pharmaceutical market speculation (Peterson 2014), and how the capacity to imagine connections between unseen things informs theories of power among disenfranchised working class communities (Lepselter 2016).

What strikes me about these works, in light of the chapters in this volume, is that those ethnographers did not create a binary between what is "only" imaginary or speculative versus what is "really" enacted. In all of these works, scholars cultivate an embodied sensibility for connections in being, space, and time that collapse what-is and what-could-be. This requires a disposition that is not just analytic but contemplative, where "to contemplate" signals the way that someone "creates a space for observation" that unbounds inner and outer, here and there, now and later, near and far.

Contemplation, with its history as a religious or sacred technology, also indexes a position of perception that holds all *modes of being and perceiving to be real and even divine*. Nowhere in this volume is this clearer than in chapters by Ojani, Chinigò and Nieber, Dovey and Potts, Reid, and Kim.

In Chapter 3, Chakad Ojani's work on the relationship between sci-fi and infrastructure building shows how the presence of "alien elements" in down-to-earth discourses about seemingly "grounded" processes like environmental governance reveal extraterrestrial dimensions of contemporary environmental governance. In Sweden, both the government and Indigenous peoples rely on orbital data from satellites to make decisions about how to manage processes like animal husbandry and mining. He realized that he wasn't going to apprehend the politics of spatial governance in Sweden by assuming Swedish "land" to be a surface space. He is inspired by Steven Shaviro's description of science fiction as a practice that "consider[s]" new social containers based on the "*virtual* dimensions of existence" (this volume, 45). Doing fieldwork, he had to fully contemplate the ways that groups of scientists in the city of Kiruna understood their city as three dimensional – as including the underground and the orbital overground, in a way that echoed (Sami) Indigenous understandings of space as transspatial.

Chinigò and Nieber, in Chapter 4, work within the super-tangible and grounded domain of infrastructure building, but they pay attention to the cosmic poetics that can be understood to "emanate" from astrophysical structures (this volume, 59). Chinigò and Nieber detail the ways in which a radio telescope array that connects South Africa and Madagascar creates a new notion of "Africa" as a symbolic portal that both receives and transmits hopeful futures across time and space. Building on Larkin's notion that infrastructures are poetic as well as political, the authors ask their readers to contemplate, with their interlocutors, a new 3-dimensional cosmography of African futurity regardless of whether a better future ends up materializing. In this case, infrastructure has a cosmopoetics that scales not just before but beyond materialization, into a field that re-imagines terrestrial spatializations.

In Chapter 8, filmmakers Dovey and Potts confront the ethnographer's interior engagement with outer spaces head on: how do you make a film about "there" without "being there"? Their answer is that one has to develop a mindset of "paying attention" (this volume, 128) that doesn't create a separation between what is imagined and what is real. They speak of the problem of acknowledging bodies as "in one place" with minds that can "still roam freely." They argue that "art does not ask you to take a position on something" – in fact, if you're trying to understand space-focused interlocutors, you have to engage in "close listening" to archives of, for example, outer space imagery in ways that amount to being able to, for example, "[feel] a deep reverence for the Moon" (this volume, 136). This sentiment indexes the capacity of contemplation as a way to diminish separation between all forms of material being, be it human or otherwise. The editors of this volume described their position to me as a "meditation on [the authors'] experience of engaging in a creative visual ethnography as a way of doing an ethnography of outer space futures as they unfold."

Lauren Reid's work, in Chapter 10, brings outer space ethnography face to face with contemplative practices. Situated within ufologist (believers in extraterrestrial spacecraft visitations on earth) groups in Thailand, Reid shows how outer space-focused practices are metaphysically positioned in the intersection of technical and religious views of reality. She is interested in how social groups come together to "make sense of unfamiliar realities" (this volume, 176) and what those practices might offer ethnographers tasked with the same problem. Her resulting fieldwork data assemblage method is as contemplative as the practices of her Buddhist interlocutors. She proposes an "ethnographic practice that engages with multiplicity by perpetually reconfiguring components." But this isn't just social scientific bricolage. It is a mode of perception. She concludes her piece with this invocation: "I advocate for thinking with a general metaphysics characterised by impermanence, decentredness, interdependence, and constant emergence" (this volume, 175–176). In this contemplative mode, Reid changes the definition of ethnography – to a practice that is constantly in a state of reconfiguration.

In Chapter 11, Hae-Seo Kim's work on the role of shamanic divination in South Korean "Space Age" makes an intervention not only into ethnographic methods but into ethnographic consciousness. She traces shamanic astrological practices from local space science and technology settings into the highest levels of government. Her work shows that "shamanistic and scientific practices co-constitute" South Korean modernity. However, Kim does not work with the dominant Western ethnographic definition of what space is. Instead, she helps her readers focus on "how outer space is understood and lived in multiple ways in different cultural contexts" (this volume, 192).

Her method for doing this is to watch how diviners like Shaman Park create spaces to "contemplate" what spirits and ancestors wish for people living in the orbit of the new satellite business world (this volume, 189). As she contemplates, observes, and participates in the outer spatial synergies of divination and aerospace technology, she shows how outer spaces are conjoined in South Korea, from birth to death. Shamanic and scientific cosmologies form a cosmology multiple, infusing politics, infrastructure building, and the very definition of South Korean space writ large.

Sensibilities as method

Kim's work calls for a shift in consciousness about what "space" is definitionally, and within anthropology in particular. Her analysis joins those of the others in this book, all of which offer new ways to sense the presence of outer spaces within all social spaces. As the ethnographers in this book seek to multiply positionalities and to contemplate the multiplicity of realities, they move beyond notions of "outer space" as somehow something other than social space.

References

Anderson, Benedict. 1983. *Imagined Communities: Reflections on the Origin and Spread of Nationalism*. London: Verso.

Battaglia, Debbora. 2006. *E.T. Culture*. Durham, NC: Duke University Press.

Berry, Maya J., Claudia Chávez Argüelles, Shanya Cordis, Sarah Ihmoud, and Elizabeth Velásquez Estrada. 2017. "Toward a Fugitive Anthropology: Gender, Race, and Violence in the Field." *Cultural Anthropology*, 32(4): 537–565.

Casumbal-Salazar, Iokepa. 2017. "A Fictive Kinship: Making "Modernity," "Ancient Hawaiians," and the Telescopes on Mauna Kea." *Native American and Indigenous Studies*, 4(2): 1–30.

Charbonneau, Rebecca A. 2021. "Imaginative Cosmos: The Impact of Colonial Heritage in Radio Astronomy and the Search for Extraterrestrial Intelligence," *American Indian Culture and Research Journal*, 45: 71–94. Special Issue "Settler Science, Alien Contact, and Searches for Intelligence," eds. Kim Tallbear and David Shorter.

Crossa, V. 2015. "Relational Positionality: Conceptualizing Research, Power, and the Everyday Politics of Neoliberalization in Mexico City." ACME: An International Journal for Critical Geographies, 11(1): 110–132. https://doi.org/10.14288/acme.v11i1.921

Goodyear-Ka'ōpua, Noelani. 2017. "Protectors of the Future, Not Protestors of the Past: Indigenous Pacific Activism and Mauna a Wākea." *South Atlantic Quarterly*, 116(1): 184–194.

Hobart, Hi'ilei Julia. 2019. "At Home on the Mauna: Ecological Violence and Fantasies of Terra Nullius on Maunakea's Summit." *Native American and Indigenous Studies*, 6(2): 30–50.

Hoeppe, Götz. 2012. "Astronomers at the Observatory: Place, Visual Practice, Traces." *Anthropological Quarterly*, 85(4):1141–1160.

Khurshid, Ayesha. 2017. "Does Education Empower Women? The Regulated Empowerment of Parhi Likhi Women in Pakistan." *Anthropology & Education Quarterly*, 48(3): 252–268.

Lempert, Willi. 2014. "Decolonizing Encounters of the Third Kind: Alternative Futuring in Native Science Fiction Film." *Visual Anthropology Review*, 30(2):164–176.

Lepselter, S. 2016. *The Resonance of Unseen Things: Poetics, Power, Captivity, and UFOs in the American Uncanny*. Ann Arbor: University of Michigan Press.

López Caballero, Paula. 2021. "Inhabiting Identities: On the Elusive Quality of Indigenous Identity in Mexico." *Journal of Latin American and Caribbean Anthropology*, 26(1): 124–146.

Messeri, Lisa and Valerie Olson, forthcoming. "Spaced Out" in *Introduction to Outer Space Studies*. Praet, I. et al., eds. Oxford: Oxford University Press.

Mirmalek, Zara. 2020. *Making Time on Mars*. Cambridge, MA: MIT Press.

Murphy, Keith. M. 2005. "Collaborative Imagining: The Interactive Use of Gestures, Talk, and Graphic Representation in Architectural Practice." *Semiotica*, 156(1/4): 113–145.

Olson, Valerie. 2023. "On Paradoxical Ground: Refielding Social Studies" in *The Routledge Handbook of Outer Space Studies*, Salazar, J, Castaño, P and Gorman, A. eds. London: Routledge.

Peterson, Kristin. 2014. *Speculative Markets: Drug Circuits and Derivative Life in Nigeria*. Durham, NC: Duke University Press.

Peterson, Kristin and Valerie Olson. 2024. *The Ethnographer's Way: A Handbook of Multidimensional Research Design*. Durham, NC: Duke University Press.

Reynolds, Jennifer F. and Marjorie Faulstich Orellana. 2009. "New Immigrant Youth Interpreting in White Public Space." *American Anthropologist*, 111: 211–223. https://doi.org/10.1111/j.1548-1433.2009.01114.x

Samatar, Sofia. 2017. "Toward a Planetary History of Afrofuturism." *Research in African Literatures*, 48(4): 175–191.

Smiles, Deondre. 2020. "The Settler Logics of (Outer) Space. Society and Space." www.societyandspace.org/articles/the-settler-logics-of-outer-space

Star, Susan Leigh. 1999. "The Ethnography of Infrastructure." *American Behavioral Scientist*, 43(3): 377–391.

Thompson, Daniel. 2017. "Scaling Statelessness: Absent, Present, Former, and Liminal States of Somali Experience in South Africa." *PoLAR*, 40(1): 86–103.

Valentine, David. 2007. *Imagining Transgender*. Durham, NC: Duke University Press.

Yusoff, Kathryn, Kerry Holden, and Casper Laing Ebbensgaaard. 2022. "Planetary Portals: 'Dreaming in Continent' from East London to the Cape in the Colonial Praxis of Emergence and Extraction." https://mediathek.berlinerfestspiele.de/en/gropius-bau/journal/planetary-portals

14 … a response

Victor Buchli

I

By way of a 'response' to this wonderful collection I want to start off with a well-known and frequently cited quote from Martin Heidegger and the expression of his horror at what seemed to him to be the end of the world in 1966, after viewing the first images of the Earth from the Moon. The last few lines, however, are rarely quoted apropos the poetic. I want to consider here the end of the world, as Heidegger saw it, the role of the poetic, as the philosopher of space Kelly Oliver goes on to reprise (2014), and the contours of a distinctive methodology and theoretical orientation that emerges from the contributions of this volume.
But first,

> We do not need atomic bombs at all [to uproot us] – the uprooting of man is already here. All our relations have become merely technical ones. It is no longer upon earth that man lives today. Recently I had a long dialogue in Provence with Rene Char – a poet and resistance fighter as you know. In Provence now, launch pads are being built and the countryside laid waste in unimaginable fashion. This poet, who certainly is open to no suspicion of sentimentality or of glorifying the idyllic, said to me that the uprooting of man that is now taking place is the end [of everything human], unless thinking and poetizing once again regain [their] nonviolent power.
>
> (Heidegger 1966, 56, Richardson translation of *Der Spiegel* interview)

It is to this injunction to attend to the poetic and its non-violent power that I wish to turn to in Kelly Oliver's work. Oliver attends to Heidegger's injunction apropos the poetic to posit a distinctive ethical disposition that emerges through the radical incommensurability of worlds. This

DOI: 10.4324/9781003437956-14

incommensurability forms a distinctive strain in the works presented here and offers a correspondingly distinctive method and ethical disposition towards the expansion of human endeavours into outer space as well as their integration into terrestrial societies.

As the various contributions in this volume attest, in the ways they attend to the radical particularity of emergent space worlds, "ethics requires us to dwell in the undecidable space of the impossibility of knowing what is right and yet being obligated to do it nonetheless" (Oliver 2014, 131). In short, there is a radical incommensurability to these emergent space worlds that we witness here and that can only be apperceived through the efforts of the editors to collect and bring together these partial emergent worlds in critical apposition. The operation in its indeterminacy, as Oliver notes, is a "certain *poetic* world-making [...] as a counterbalance to sovereign world-building" (Oliver 2014, 122). Thus, as Oliver notes, such a poetics of radical incommensurability requires a novel ethical disposition. One must attend to "the poetic *as-if*, as if the world has constantly renewable meaning. Neither the poet nor the poem is sovereign" (Oliver 2014, 129). As Oliver suggests, this engenders an ethical responsibility – not in relation to a given dominant sovereign order as morality would dictate – but by being, as she calls it, "response-able" outside of sovereign moral orders as a form of hospitality that bears witness to these emergent worlds "that both assumes and creates the possibility of response-able cohabitation" (Oliver 2014, 133). Here we can see in this volume how that 'response-ability' works as they address and accommodate in appostion to one another a variety of incommensurate and local emergent 'cosmoses.' The contributions to this collection thus provide conceptual and methodological approaches to help engender these 'response-abilities' and with that show how novel incipient ethical dispositions might emerge without a sovereign order and within which multifarious 'inevitabilities' arise.

Thus the many fragmented worlds that are both extremely local and extremely cosmic do not necessarily speak to each other directly: "We must act *as if* we are creating the world anew, reinventing the wheel. And, yet, to do so ethically demands that we do not create as lords or masters of the world or earth, but rather, as care-takers who acknowledge our dependence on other" (Oliver 2014, 134). Thus the contributions of this volume are able to 'address' one another in this hospitable tone through the montage of critical apposition. Here, genealogies of various forms of the 'inevitable' future and space, tying together extreme localisation and extreme space, offer a new take on the extreme and the sublime. More critically, we see the methodological importance of deep time perspectives that attend to the *longue durée,* because emerging neo-colonial and neo-evolutionary narratives demand this change in method, theoretical orientation, and training. The contributions here naturally point to this methodological extension.

The chapters empirically describe how the fragmentation of the 'inevitable' outside of any sovereign narrative enables multiple voices and worlds that articulate disparate 'inevitabilities' in various locales and their attendant 'milieux' (Johnson, Chapter 7, this volume). Such a fragmentation goes hand in hand with the advance of commercial space and the fragmentation of space efforts away from state centred enterprises. The two go hand in hand, and late capitalist practices adroitly incorporate these dynamics. Yet, as Hirokazu Miyazaki suggests, one can be harnessed to the other with novel effects, as we see here. Such a fragmentation produces a plethora of "and/or" and "not this/not that" that are at the heart of Elizabeth Povinelli's discussion of incipient, novel, and radical forms of life. These Povinelli understands "as a series of quasi-events that provide the preconditions in which some new social content might be nurtured" (Povinelli 2011, 191). The ethnographic method, and the partial emergent worlds presented herewith, work to enter these worlds along their sublime axis of the extremely local and extremely cosmological, to constitute 'response-ably' one another and wider processes of equitable, more just, and ethical "response-able co-habitation" (Oliver 2014) and the multiplicity of worlds these expansions and innovative ethical imperatives locally and collectively engender. Such an emphasis on the poetics of incommensurability over sovereign narrative promises to relieve Heidegger's melancholia in the face of the destruction of '*the*' world as "thinking and poetizing once again regain [their] nonviolent power" (Heidegger 1966).

Of relevance to these poetics and 'their nonviolent power,' Sarah Pink has written about how methodologically, theoretically, and ethically one writes about emergent futures (Pink 2023). Most of the contributions here and most ethnographic work on outer space in general deals with the question of emergent futures and their implications for daily life in various locales. In particular, she argues for a form of 'hope' that is an imaginative and ethically inflected methodological disposition that attends to the ways in which emergent futures are integrated in actuality, in daily lives, and modes of being (Pink and Salazar 2017). Specifically, she attends to how new technologies (such as AI and remote work), technocratic 'hype' from tech proponents, and emergent futures shape human societies (Pink 2023). Pink argues against dismissing 'hype' straight away, electing instead to examine the nuanced ways 'hype' is sedimented into daily life and how local and radically divergent ways of inhabiting such new technologies produce a distinctive "otherwise" (Povinelli 2011). The 'otherwise' cannot, of course, be predicted, but nonetheless emerges in empirically available and non-arbitrary ways, as many of the contributions here demonstrate.

Pink argues for 'hope,' and I would argue its radical and poetic force as a methodological disposition of 'response-ability,' as follows:

> If we centre hope as a prism through which to understand the meaning of everyday interventions with emerging technologies, we might create shared hopes, or engage hope as an affective mode of empathy. It is through experimenting with people in everyday life that hope might become materialised and experienced in and through emerging technologies, in place of hope being projected from an external source, as the promise of technological solutionism.
>
> (Pink 2023, 51)

Pink cites Miyazaki here to further her point: "For Miyazaki, hope, as a method, 'unites different forms of knowing'" and, in a manner that I want to suggest is 'response-able' following Oliver and is not just a way of bringing together diverse knowledge and experience, "it can also perform as 'a method of radical temporal orientation of knowledge'" (Pink 2023, 49, citing Miyazaki 2004, 4–5).

This is what the contributions to this volume of course do here. They do this not simply in terms of their critical analyses of how outer space is integrated into diverse contexts, producing multiple outer spaces and different forms of worlding as these futures are sedimented within the particular and historically inflected conditions of daily life in their respective locales. But, as the archaeological metaphor of sedimentation implies, these emergent worlds are further integrated in deep historical structures that often can be dated to several hundred years. Many of the contributions of this volume naturally integrate a historical perspective into the structures that support emerging worlds. Methodologically, this becomes an ethical as well as analytical imperative, especially when considering earlier colonial legacies.

The volume itself works to facilitate a "response-ability" (Oliver 2014) to these various incipient futures that through their firm ethnographic grounding in the conditions of daily life and historical structures produce this 'diversity' – and thereby the 'otherwise' through critical apposition. It is at once a method focused on the specific scale of a given community on one hand, and also on the inherent indeterminacy of these 'futures' and how they emerge within local historical structures and against dominant technocratic "hype" (Pink 2023). This is what the comparative ethnological method does historically and to novel effect here, as these various contributions attest moving "towards instead a mode of hope which stems from what is incrementally learned and knowable through our everyday attention to the ecologies of which we are part," as Pink states succinctly following Tim Ingold (Pink 2023, 43). Hope thus becomes a resource that emerges within excess when incommensurability produces an outside – an outside that is available as a resource to produce "not this/not that" – and is in turn generative of unanticipated dispositions as is poetics wont to do. The contributions in this volume speak to these unanticipated dispositions as they emerge through the methodological ethnographic commitment to the everyday and the creative capacities of our interlocutors to generate "otherwise" and "not this/not that" (Povinelli 2011, 191).

It has been noted by figures such as Valerie Olson and Lisa Messeri that there are as many outer spaces as there are 'milieux' (Johnson, Chapter 7, this volume) that inhabit the expanding nexus of earth and outer space. One can speak to the demise of earlier 'sovereign' master narratives of outer space with the waning of the singular bi-polar dominance of national space programmes, namely between the US and the USSR/Russian Federation. As claims for outer space expand from those of Cold War superpowers to new national space programmes and commercial enterprises and to a plethora of informal and formal agents across the global north and south, a multiplicity of outer spaces emerges within this highly dynamic and expanding nexus.

I would like to suggest that this empirically evident and incommensurable multiplicity is not so much a 'problem' that needs to be resolved in relation to a given paradigm, but the very circumstances in which new conditions of ethical action are enabled and theoretically sustained. This collection, admirably put together by the editors and their contributors, by virtue of its inherent multiplicity across geographical locales and varied empirical contexts doesn't just simply 'sample' and catalogue this diversity. It is precisely the edited volume with its montage-like qualities and arrangement in critical apposition that speaks to the ways in which the gaps that emerge are productive of the very "response-able" and hospitable (Oliver 2014) conditions by which novel ethics might be seen to emerge and materially made manifest. In fact, this is precisely the technique employed by Vincent Crapanzano himself to develop his exegesis on hope, writing: "I depend on the montage, on the disquiet produced by the juxtaposition of events, images, and theories that are not easily coordinated" (2003, 4). This is not just simply a limited claim for 'otherwise' and a demonstration of 'not this, not that' in Povinelli's claims for the emergence of a novel ethics of action in the "subjunctive mode" (to reprise Battaglia et al. 2012, 1011). Rather, it is a mapping of the way anthropologists of outer space (fully cognizant of the inadequacy of this bifurcation into 'outer') 'enter the field' at various locations and thereby further constitute it through their participation, witnessing the emergence of these multiple 'outer spaces' and the often incommensurable worlds that accompany them. As Miyazaki notes, citing Crapanzano, in relation to the mutuality of the workings of hope, the distinction between analyst and subject are collapsed when "[w]e are all, I suppose, caught" (Crapanzano 2003 in Miyazaki 2006, 148) within these 'milieux' (Johnson, Chapter 7, this volume). Our entry into the field is productive of the field but also complicit in its expansion. But by virtue of our engagement, it also produces an understanding "of hope as a general principle that unites analysts and their subjects" (Miyazaki 2006, 148). This is not the case of 'getting it right' or finding an 'adequate language' to describe this emergence, but instead, by bearing witness to the very cacophonous and inherently irreconcilable terms by which these 'cosmoses' emerge, and the conditions created whereby one is able to be "response-able" and hospitable (Oliver 2014) to one another and the novel ethical demands that arise therein. One is not 'responsible' as Oliver notes in relation to the imperatives of a given sovereign morality or dominant moral order but one becomes instead, as she states, 'response-able.' It is to be 'response-able' to one another that the works in this volume herein address these multifarious incommensurate and local emergent 'cosmoses.' It is precisely these 'response-abilities' that emerge within the gaps through critical apposition, that I would like to discuss here in relation to the contributions. Throughout the works presented here, certain conceptual and methodological operations are proposed that speak to the way in which novel incipient ethical dispositions emerge without a sovereign order and within which multifarious 'inevitabilities' arise.

II

Diverse poetics of incommensurability are at play in the production of multiple local inevitabilities in how terrestrial locales and their space milieux (Johnson, Chapter 7, this volume) emerge. On their own they represent partial worldings, but together they reveal the range of techniques whereby these multiplicities emerge, and suggest how we as students of the anthropology of outer space are able to be 'response-able' to the workings of these poetics. Two key yet related methods seem to emerge which enable this poetics of 'response-ability' towards these emergent 'inevitabilities.' One is based on a historiographic emphasis on the *longue durée*, the other on the representational techniques of bricolage and montage.

Longue durée

The emphasis on the *longue durée* in a number of the contributions operates here to provide a means by which to understand the deep historical structures that enable a seemingly 'inevitable' emergence of a given space future. Ojani, in Chapter 3, notes how the "thickening infrastructural relations between Earth and space might be reshaping the domains drawn together" (this volume, 51). As Ojani observes, the imaginative extension of this vertical axis both into outer space and underground produces the 'inevitable' in Kiruna, a Swedish subarctic city, "as a by-product of off-earth mining," where the launch site emerged as a result of a local history of mining: "the city's relation to mining served to envisage human settlement in space" (this volume, 53) creating an imaginative axis in deep time that links non-arbitrary historical mining practices with incipient futures.

Similarly, in Chapter 4, Chinigò and Nieber methodologically draw the ethnographer and historian together and extend the axis of analysis of the African Square Kilometre Array (SKA) in relation to expanding futures and sedimented colonial pasts. The SKA adds to historically sedimented colonial "civilising missions" and erasures to produce a futurity of hope that facilitates a regionally specific manifestation of the cosmos whereby "this 'Africa' elicits hope, desires, and potentials" (this volume, 72). Here, this particular manifestation of the vertical axis in a 'response-able' manner, both temporally and spatially, can attend to both the vastness of outer space, the immanent histories of a small town, and neighbouring 'African' aspirations.

Osbourne, in turn, in Chapter 5, addresses a single meteorite on display in Greenwich and posits an alternative interpretive historical axis where geographically distributed genealogies of colonialism and extractivism congeal. In particular, Osbourne notes how alien substance is conscripted into imperial cosmologies, and how the sensuous intimacies of the exhibit facilitate this 'inevitable' axis of resource exploitation at the very point where such an axis bifurcates the entirety of the world at the Greenwich meridian. This Osbourne addresses in sensuous terms which suture these seeming incompatibilities: "how poking the crevasses of a meteorite on a Sunday morning might intimately connect museum goers to past and future imperial projects of extractivism, and beyond our planet" (this volume, 81). Here, seemingly naïve touch seeks to reconfigure a cosmological orientation that speaks to the inevitability of emerging narratives of futurity and historicity. As Osbourne notes, "[by] making the alien familiar, extraterrestrial and earthly rocks become analogous, collapsing otherwise incommensurable scales of separability, and incorporating meteorites into a geological fold" (this volume, 88).

Tereshin and Sivkov in Chapter 6 reprise the theme of amplifying the significance of the local against the backdrop of the infinite universe, exploring in many ways the way in which I would argue the Kantian sublime produces an incipiently radical new sensorium and form of personhood. Such regional museums enable intimate sensuous encounters with space artefacts in a distinctly local context. These encounters enable an 'inevitability,' as other contributions here have shown, in terms of a given habitation of the cosmos and expansion of the terrestrial nexus outwards into space. These encounters take place in regions that traditionally served as historical 'terrae nullius' which "paved the way not only for the steady progress of the Soviet Union but also, by extension, the emergence of the Soviet space industry" (this volume, 93). Tereshin and Sivkov note how the sovereign narratives of space habitation both of the USSR and the Russian Federation are fragmented by these local museums in distinctive ways and in particular in terms of assembling a highly local cosmogonic "verticality." As they write, the affective localisation of cosmic achievements produces a radically local 'inevitability': "The transition from the history of space exploration to the local seems almost seamless" (this volume, 103). They argue that such local museums amplify "unlike other places" (this volume, 105) in that they further fragment against any form of collective sovereign representation – it is not a fragment of something larger or fractal, it is in fact '*the*' entire universe itself.

Similarly, Johnson, in Chapter 7, employs the *longue durée* of historical singularisation and with that the 'inevitable' emergence of a Mexican futurity in space. Johnson notes that, much like in the earlier Russian example of Tereshin and Sivkov, as the home of a Mexican astronaut, the Mexican state of Guerrero "is in space" according to her interlocutors (this volume, 119). She develops the concept of the 'milieux' to describe the multifarious ways in which the local and cosmic emerge in relation to one another within this *longue durée* and how they are entered and complicitously constituted by the ethnographer that is at once "response-able" (Oliver 2014) and poetically constitutive of locally inflected and dynamic space futures.

In a similar manner responding to deep historical structures, Kim in Chapter 11 speaks to the distinctive fashion in which the South Korean space programme is intimately linked to long established historical and local shamanistic and astrological cosmologies. In particular, Kim attends to the ways in which the future is secured through both traditional divinatory practices related to shamanism and astrology and to the more scientistic practices of the national space programme and its attendant industries that secure a future for South Korea in space. As Kim notes, the two seemingly paradoxical and incommensurable practices at play would seem to eliminate one another. But with young people moving in and establishing families, these traditional practices are necessary for people to secure their futures within the expanding space sector. One aids the other – enabling families and "provid[ing] counsel through understanding the placement of celestial bodies in the cosmos and [guiding] people through one's life path and temporal decision making" (this volume, 191). Thus "shamanistic and scientific cosmologies co-constitute the material and social relations that make up South Korea's 'Space Age'" (this volume, 192) and produce an 'inevitability' that is radically incommensurate with others.

Montage/bricolage

The techniques of montage and bricolage employed by some of the authors here perform novel imaginative and poetic alignments for the inhabitation of space and their local 'inevitabilities.' In fact, in Chapter 8, Dovey and Potts argue for a deliberate and purposeful fragmentation of any one sovereignty following Russell as a "dismantling [of] the universalist impulse of realist aesthetics" (Russell quoted in Dovey and Potts, this volume, 132). Here they invoke the use of montage – a well-established technique of world building and vivification since the Soviet master of the method Dziga Vertov. In the process of montage, breakdown and reassembly literally animate and vivify through their inherent destructiveness, rip apart sovereign worlds, and counterintuitively, in their own words, "intensify" in a novel fashion one's relation to cosmological objects such as the Moon. *Requiem*, their film about the eventual decommissioning of the ISS, mourns in anticipation the foreseen "death" of this world, which is "splintered and refracted" through their visual form-making and brings the "ISS down to Earth" in highly affective and emotive terms. In doing so they invoke a poetic 'response-ability': through these interventions they "shift people's epistemic certainty – *just a tiny bit*" (this volume, 142). and thereby constitute a novel set of incipient relations.

Similarly, in Chapter 9, Jeevendrampillai and Fortais, an ethnographer and artist respectively on a remote Scottish island/Mars analogue, produce a distinct habitation of both Earth and Mars through the technique of bricolage. A dynamic emerges that vacillates – at once denying one and the other – while constituting something novel and distinctive within these vacillating cancellations. Here 'bricolage' produces a distinctive 'island' in the fullest sense of the word, because it denies all other meanings through its alternating multiplicities, and constitutes a vertical axis of habitation and sensorial re-attunement that must be 'response-able' to the new, hybrid, and dynamic subjectivities that emerge. As they note:

> For us, bricolage is not about building towards an a priori imagined finished thing, but rather creatively fashioning joints/connections thereby embedding oneself into the process of making art and anthropological knowledge [...] This served to build a collective sense of being in a particular place, Earth, *Mars* or somewhere in between (this volume, 160).

Thus an imaginative 'world' is forged out of alternating erasures that produce an incipient 'inevitability' that a distant world has to be inhabited in relation to ours.

In Chapter 10, Reid develops a technique of examination that destabilises normative logics as a resource: "that engaging with multiplicity ... offers a generative approach to apprehending other worlds and possibilities" (this volume, 166). Here, Thailand becomes the site where a "fledgling yet rapidly developing field of space presents a de-centred context from which to observe the flows of space-scientific practices and knowledges in relation to various local ways of knowing and being in the cosmos" (this volume, 166). Like the juxtapositions produced through bricolage and montage, they engender a compelling resource with which to critically confront "normative logics" (this volume, 175). Engaging in a method of "ethnographic conceptualism" (following Nikolai Ssorin-Chaikov) to "generate ethnographic situations" through the juxtaposition of diverse perspectives and sites related to the inhabitation of space, at the hands of all workshop participants they are 'bricolaged' to enable such destabilisations. The method of bricolage here offers "a generative approach to asking new questions and inhabiting worlds otherwise" (this volume, 174).

Finally, Timko and Korpershoek in Chapter 12 speak to the productive capacity of gaps such as those formed through bricolage and montage to illustrate how leaving 'space' undefined or flexible helps to bring to the fore divergent understandings of space on the ground in the discussion of their respective field sites in southern California and French Guiana. Through deep ethnographic immersion and comparative juxtaposition, both studies speak to the vertical axis of history, place, and histories of colonial expansion to consider the multiplicitous ways in which space is inhabited in these distinct yet similar sites with complex local histories. As they state, " 'outer space' does not exist as separate from Earthly power practices," and as they demonstrate from their deep histories, they are most resolutely local and particular 'worlds' of inhabitation though relatable through their similar technocratic institutions. Their work results in an imaginative and methodological sensibility, where "[l]eaving space blank allows for these different configurations to a rise from how 'outer space' is experienced, practised, and/or imposed on people. The methodological challenge of studying 'outer space' is met by looking around, behind, and forward as well as looking up" (this volume, 205).

III

Thus the works presented here advance an operational "aesthetics of emergence" (Miyazaki 2004, 137–9) with their methodological and interpretive techniques by which an emergent mutuality and 'response-able' approach might be constituted through specific operations such as a critical constitution of the *longue durée* and the use of bricolage and montage to understand the emergence of terrestrial communities expanding into space. Collectively through their presentation in apposition to one another, they also speak to the different ways past structures in various registers "replicate" (Miyazaki 2004) anew and in radically distinctive and unexpected fashion across these communities that in turn reconfigure conventional understandings in a mode that is "subjunctive" (Battaglia et al. 2012) and "hopeful" (Crapanzano 2003; Pink 2023; Miyazaki 2004). They suggest how one might realise Miyazaki's injunction to attend to how "hope is inherited from the past, and the pull of hope in the present derives from anticipation of fulfilment contained in that past hope." (Miyazaki 2004, 139). And as regards these highly localised, contingent, and emergent futures, how "the effort to maintain prospective momentum entails an effort to replicate a past unfulfilled hope on another terrain" (Miyazaki 2004, 139). Here, however, that terrain is not figurative but, literally and materially, no longer of 'terra.'

More immediately, the contributions speak to the way ethnographers 'enter' these sites and co-constitute them to suggest novel and hopeful understandings of these emergent communities and their futures. These contributions through their "response-ability" (Oliver 2014) perform the work of "increasing *sensitivity*, increasing responsiveness to the needs of a larger and larger variety of people and things" (Rorty 1999 [1994], 81). In his essay 'Ethics without Principles,' Richard Rorty urges this as the way forward past the 'post-modern' malaise of the social sciences (see also Miyazaki 2006 and Pink 2023), as part of the "the process of adjustment which is also the process of recreating human beings" (Rorty 1999 [1994], 81). And keeping with this injunction, the collection methodologically 'widens' empirically through critical apposition to achieve the goal of "wider and wider sympathy" (Rorty 1999 [1994], 82) and thereby, I would argue, in highly poetic and rigorously empirical terms 'widens' the purview of "the Whitmanesque and Whiteheadian romance of unpredictable change" (Rorty 1999 [1994], 88). We can see how in very precise methodological and technical terms one can more effectively and ethically engage with such emerging space futures and modes of inhabitation in the material conditions of actuality, as the editors of this volume have so 'hospitably' and effectively presented here. Through the collective apposition of these works, montage-like, one provides a view through these emergent operational aesthetics (Miyazaki 2006) and methods onto how empirically and theoretically we might make sense of these emergent and divergent worlds unbound by 'terra,' as we conventionally know it. And one points at how we might be able to apprehend them and inhabit them more hopefully and 'response-ably' and bear witness to "the need to create new ways of being human, and a new heaven and new earth for these new humans to inhabit" (Rorty 1999 [1994], 88).

References

Battaglia, D. et al. (2012). Introduction: Extreme Limits and Horizons in the Once and Future Cosmos, *Anthropological Quarterly*, 85(4): 1007–1026.

Crapanzano, V. (2003). Reflections on Hope as a Category of Social and Psychological Analysis, *Cultural Anthropology*, 8(1): 3–32.

Heidegger, M. (2010 [1966]). "Only a God Can Save Us": The Spiegel Interview. In T. Sheehan, ed., *Heidegger: The Man and the Thinker*, London: Transaction Publishers.

Miyazaki, H. (2004). *The Method of Hope*, Stanford, CA: Stanford University Press.

Miyazaki, H. (2006). Economy of Dreams: Hope in Global Capitalism and Its Critiques, *Cultural Anthropology*, 21(2): 147–172.

Oliver, K. (2014). The Poetic Axis of Ethics, *Derrida Today*, 7(2): 121–136.

Pink, S. (2023). *Emerging Technologies/Life at the Edge of the Future*, London: Routledge.

Pink, S. and J.F. Salazar (2017) Anthropology of Futures: Setting the Agenda. In J.F. Salazar et al., eds, *Anthropology and Futures: Researching Emerging and Uncertain Worlds*, London: Bloomsbury Academic.

Povinelli, E. (2011). *Economies of Abandonment: Social Belonging and Endurance in Late Liberalism*, Durham, NC: Duke University Press.

Rorty, R. (1999 [1994]). Ethics without Principles. In R. Rorty, ed., *Philosophy and Social Hope*, London: Penguin Books.

Index

Note: Page numbers in *italics* refer to figures, Endnotes are indicated by the page number followed by "n" and the note number e.g., 111n5 refers to note 5 on page 111.

Kolibri 360 GmbH
Ferdinand-Jühlke-Straße 7
99095 Erfurt, Deutschland
produktsicherheit@kolibri360.de